The

COMPOST TOILET HANDBOOK

by Joseph C. Jenkins

Also by Joseph C. Jenkins: *The Humanure Handbook*, 4th edition

ISBN 13: 978-1-7336035-1-5

Library of Congress Control Number: 2021905682

First Printing May 2021. Printed in the USA.

Published by Joseph Jenkins Inc.

143 Forest Ln, Grove City, PA 16127 USA

Ph: 814-786-9085 • Web: JosephJenkins.com

Please address all retail and wholesale book orders to our distributor:

Chelsea Green Publishing, Inc.

85 North Main Street, Suite 120, White River Junction, VT 05001

Tel: 802.295.6300; Fax: 802.295.6444

Orders: 800.639.4099; Office Hours: 9am – 5pm (EST)

ACKNOWLEDGEMENTS

Special thanks to Samuel Autran Dourado and Alisa Puga Keesey for the extraordinary compost toilet projects they created on several continents that the author reviewed and documented, and for contributing photos and data. Also, thanks to Patricia Arquette, Ole Ersson, Linn Harding, Joseph Lawrence Lacha, Jerry Erbach, Otgonbayar Galbadrakh, Samson Muhalia, Tom Griffin, Corrine Coe-Law, Sasha Kramer, Ruth Dykyj, Martijn Nitzsche, Troels Kolster, Lucho Jean, Ray Abbassian, Phoebe Jenkins, Jeanine Jenkins, Marsha Tillia, Sunshine Jenkins, Maxine Hoffmann, and Joseph J. Jenkins for their contributions to this book. There are so many other people around the world the author has had the pleasure of working with, far too numerous to mention, who also helped in the creation of this book, one way or another, whether they know it or not. Although their names are not listed here, they are remembered with great respect and gratitude.

Cover design by:

Kelsey Brown and Elena Reznikova, DTPerfect.com

Proofreading: Eileen M. Clawson

Printed in the USA on FSC certified paper

Photos are by the author unless otherwise indicated.

Cartoons on pages 22, 104, 213, and 238 are adapted from original artwork by Tom Griffin.

The Compost Toilet Handbook

TABLE OF CONTENTS

If you know an authority who needs this book, the author will provide a free copy, no questions asked. Just email the publisher the mailing address (US only) for a paper copy, or the email address for foreign residents, who will be provided a digital download.

INTRODUCTION

This book is a companion book to *The Humanure Handbook,* 4th edition (Joseph Jenkins Inc., 2019). Its purpose is to provide basic information about how to make compost from toilet material in a variety of situations, locations, and cultures. It is lighter on text and heavier on illustrations relative to *The Humanure Handbook*, in order to create a book that is easier to translate into foreign languages. *The Humanure Handbook* was on its 21st foreign translation at the time this book was being written. Obviously, there is a world-wide interest in the constructive recycling of human excrement, especially where water toilets are non-functional or impossible.

The Compost Toilet Handbook is intended for readers who are interested in learning how to make and use compost toilets, for everyday use, for emergency use, or for specialized use such as in schools, orphanages, remote camps, villages, or off-grid buildings. It focuses on the nuts and bolts of compost toilet construction, use, and practice, as well as compost management. It is based on the author's over four decades of continual compost toilet use in northwestern Pennsylvania, United States, including 15 years of travel teaching compost toilet methodologies worldwide.

Regarding repetition in this book, remember that the book is available for download, chapter by chapter. Hence, people may download just one chapter or another, not the entire book. Important points are therefore repeated in some chapters. Many scholarly references for information presented in this book can be found in *The Humanure Handbook,* 4th edition, which is also available for download, chapter by chapter.

This is a practical niche book relying heavily on color photos and illustrations. Although the 161 color pages have made this book expensive to print, a picture is worth a thousand words, and the cost is worth it.

To understand compost toilets, one must understand composting. If you want to collect and compost "toilet material" in your home, at your event, or for your village, instead of producing sewage, then the practice of composting should be studied and understood. Hopefully, this handbook will provide a fundamental basis for that understanding.

Opposite: Household compost toilets were established by WeltHungerHilfe in Namalu county, Karamoja, Uganda, Africa.

Granddaughter playing in the author's garden.

PART ONE
Chapter One

EVERYONE NEEDS A TOILET

If you drink or eat in the morning, you know that a toilet will soon be needed. You will want to have a private, clean, pleasant, and convenient toilet nearby. But what if you don't have running water, drains, pipes, or even electricity? Will you have to trudge outdoors, in all kinds of weather, any time of the day or night, to relieve yourself behind a tree or in a hole in the ground? And possibly drag small children with you? Or do you have to use a toilet that simply discharges raw sewage into the environment?

Not if you have a compost toilet. A compost toilet can be located anywhere, indoors or out. No pipes, drains, running water, vents, chemicals, or electricity are needed. Your compost toilet can be right beside your bed, or beside your handicapped grandmother's bed. You can have it in your office, garage, shop, barn, or camp or in a tent, a mansion, a palace. If you know how to use it, it will be odor-free. The toilet will be part of a waste-free system that makes compost, and gardens.

Common definitions of the word "toilet" include "*a fixture that consists of a water-flushed bowl and seat used for defecation and urination.*" Or "*a bathroom fixture consisting of a bowl, usually with a detachable, hinged seat and lid, and a device for flushing with water, used for defecation and urination.*" Or "*a large bowl with a seat, or a platform with a hole, which is connected to a water system for when you want to get rid of urine or feces from your body.*"

No mention is made in any of these dictionary definitions, and undoubtedly in scores of others, of a toilet *not* being something connected to a water system and used for disposal. However, a toilet is any device or system, even a hole in the ground, that is used for the collection of body excretions for the purpose of disposal, or *recycling*.

The insistence that toilets require water for flushing is probably why two and a half billion people in the world have no toilets at all. The excretions of these people are deposited into the environment wherever the person can find a suitable place to squat. They can't deposit their "waste" into

Child drowns in school pit latrine, again

15 March 2018 - 14:45

Pit latrine collapses, two children die

PROTECT SCHOOL CHILDREN FROM DROWNING IN PIT LATRINES!

The Unending Tragedy Of A 5-Year-Old Who Drowned In A Pit Latrine

Another child dies in a pit latrine

Athandiwe Saba Bongekile Macupe Raeesa Pather 16 Mar 2018

Mom blames school for boy's toilet death

Arua pupil dies in school pit-latrine

Five-year-old learner drowns in a pit toilet

Typical pit latrine in Tanzania, Africa.

a bowl of drinkable water, as the civilized world would do, as there are no flush toilets to be found. When a flush toilet is being used in low-income communities, the sewage may discharge directly into a ditch, stream, or backyard because permitted sewage treatment systems are unaffordable.

Billions of humans risk disease through inadequate sanitation systems that expose them to human excrement in their outdoor living space, in their drinking water, on their fruits and vegetables, and through recreational water activities. Nearly a billion people still practice open defecation; they squat outside in a field or in their yard or garden, every day. Still many more only have a hole in the ground for a toilet. These "pit latrines" are located away from the living area because they're full of excrement, the stench is awful, they breed large fly populations, and small children can fall into them; some even die there. Pit latrines also pollute groundwater, a primary source of drinking water throughout the world.

Providing comfortable, secure, convenient, odorless, indoor, sanitary, ecological toilets for so many people is a dilemma that has puzzled developers and sanitation workers for generations. Yet human excrement is a recyclable resource, and when thought of as something of value rather than as a waste material, an entirely new direction in sanitation is revealed.

Public toilets in Kolkata, India, offer little in the way of privacy.

Pit latrines, such as this one in Mongolia, can be very dangerous for young children who fall into them and sometimes die there.

Humanure

Nothing that is recycled should be referred to as "waste." Recycled human excrement (feces and urine), known as “humanure” (not “human waste”), provides food for microorganisms. Microbes consume humanure, then convert it into soil. Although humans find their own excretions disgusting, thankfully microbes don’t. When *composting* is used as a sanitation system, sewage can be greatly reduced or entirely eliminated, as can diseases and parasites associated with fecal contamination of the environment; toilets can be located indoors where they’re safe and convenient; and they can be odor-free. The resulting compost is suitable for growing food.

In 2018 the World Health Organization declared that people have a right to water and sanitation, and that after decades of neglect, the importance of access to safe, sustainable, and well-managed sanitation systems for everyone, everywhere, was recognized as an essential component of universal health coverage. The human right to sanitation entitles everyone to private, dignified sanitation services that are physically accessible, affordable, safe, hygienic, secure, and socially and culturally acceptable. The United Nations agrees that sanitation facilities must be available and affordable for everyone, even the poorest. When people don’t have access to functioning water toilets, compost toilets can provide an alternative.

A “compost” toilet is a toilet designed to collect the organic material that typically goes into any toilet, namely feces, urine, toilet paper, and so on, for the purpose of recycling it by composting. We will refer to this organic material as “toilet material.” Toilet material can also include other human excretions such as vomit and menstrual blood. When using a compost toilet, all the toilet material is collected and recycled. No waste goes into a compost toilet, and none comes out. Organic material goes into a compost toilet system. The finished product is compost.

What Is Compost?

Compost, like agriculture, is a human creation.[1] You will not find compost in nature unless humans created it. You will not find a field of wheat,

[1]An astute reader has pointed out that the Australian Brushturkey makes compost piles and uses them to incubate its eggs!

Above: This is a pit latrine at a girl's high school in Kwale, Kenya, Africa. School children complain of slipping and getting their leg caught, and maybe broken, in the squat hole. **Opposite**: The author, in 2018, is cementing a pit latrine hole shut to block the punishing odor and swarms of flies and to convert the space into a clean compost toilet stall at a school in Uganda, Africa. The contents of the pit latrine are teeming with maggots. The resulting odorless, fly-free, compost toilet is shown on the next page. The toilet receptacle simply slides out the front of the toilet cabinet when it needs to be "swapped out." This is a WeltHungerHilfe project in collaboration with GiveLove.org.

Photos by Samuel Autran Dourado.

an apple orchard, or a vineyard either, unless humans put them there.

Compost, by definition, has three components: (1) human management; (2) the generation of internal biological heat; and (3) aerobic organisms that thrive in the presence of oxygen. If all three of these conditions don't exist simultaneously at some point in the process, then it's not composting, and the end product should not be referred to as "compost."

To make compost, humans pile organic material, such as plant products and manures, aboveground in such a manner that microscopic organisms consume them. The finished material is sometimes referred to as "humus," "earth," or "soil," but the correct term is "compost." The process that converts the material into compost requires internal heat generated by bacteria inside the compost pile. These microscopic organisms live in the presence of oxygen and are classified as "aerobic." If dumped into a pit, the organic material is likely to become inundated by water and to become "anaerobic," or without oxygen. Anaerobic organisms do not make compost.

In 2018, the US Composting Council defined compost as *the product manufactured through the controlled aerobic, biological decomposition of biode-*

This is a pit latrine converted into a compost toilet at the Police Academy children's school in Moroto county, Karamoja, Uganda (a WeltHungerHilfe project).

gradable materials, which have undergone internal biological heating, thereby significantly reducing the viability of pathogens (disease-causing organisms) and weed seeds, while stabilizing the finished product such that it is beneficial to plant growth.

The Association of American Plant Food Control Officials approved the new definition for compost because it emphasized the pathogen removing biological heating process, differentiating it from many products often confused with compost, such as earthworm castings.

This definition helps the producers of other products, from biochar and mulch to dehydrated plant material and anaerobic digestate, to differentiate their products as *not being compost.* "Vermicomposting," as an example, is a misnomer. The correct term is "vermiculture," where earthworms eat rotting organic material and convert it into worm poop. The end product of vermiculture is not compost; it's worm castings. Vermiculture is not dominated by aerobic bacteria generating biological heat. It's dominated by red worms. Heat-producing microorganisms would kill those worms and turn *them* into compost. Worm castings are not the same as compost and should not be referred to as compost.

Most, if not all, so-called "*composting* toilets" are not toilets that make compost at all. Composting does not take place inside toilets unless internal biological heat is being generated, which is unlikely. What people incorrectly refer to as "composting toilets" would correctly be referred to as "dry toilets," or "biological toilets," even "eco-toilets." The composting industry has worked hard to train and educate people about natural processes that use microbes to reduce or eliminate pathogens. The correct use of language is important. This is particularly relevant to compost and to toilets.

The conversion of organic material by microorganisms through the composting process is nothing short of magical. When we compost toilet material, we eliminate human disease organisms, or greatly reduce them, often to undetectable levels. Although the raw application of human excrement to agricultural fields has been practiced in various parts of the world for centuries, this practice can allow disease organisms to reinfect humans. Composting, on the other hand, destroys the disease organisms, creating a friendly, pleasant-smelling compost safe for agricultural use.

Pit latrine in Mozambique, Africa.

Compost Toilets and Dry Toilets

A "dry toilet" is any toilet that doesn't use water to flush away the material collected in the toilet. A dry toilet can be a urine-diverting toilet, a chemical toilet, an incinerating toilet, a biological toilet, an eco-toilet, or any of a multitude of devices designed to collect, process, dispose of, or recycle toilet material without relying on water.

Composting is unlikely to take place inside a dry toilet because sufficient biological heat will not be generated, for several reasons. For one, the mass of the collected toilet material may be too small; for another, the collected material may be too dry due to urine separation or intentional dehydration. Too *much* liquid in a dry toilet can create anaerobic conditions with consequent unpleasant odors. Although these toilets do not make compost, they can create decayed organic material, or what's known as "septage," which has *not* been subjected to the biological temperatures of true compost and is therefore not sanitized. The septage from dehydration toilets can be used as a soil additive where the material will not contact food crops, or it can be composted in a compost bin.

One of the reasons dry toilets don't generate biological heat is that the volume of the material inside the toilet chamber is too small. Compost bins of at least a cubic meter in size have a greater potential to maximize heat generation. Composters should attempt to somehow insulate compost bins and keep some type of cover on top of the compost to protect from excessive rainfall and to insulate the pile. Chambers in dry toilets tend to be much smaller than a cubic meter, and even when they are large, it is difficult to insulate around the organic material inside the receptacle, where the toilet contents may be right up against a plastic or metal wall.

Most dry toilets are designed to dehydrate their contents. This is often achieved by "urine diversion," by either diverting the urine away from the solids at the source utilizing a toilet seat designed for this purpose, or by allowing the urine to drain away from the toilet contents, perhaps directly into the soil underneath a toilet stall. Dry toilets may also utilize electric or solar heating elements to further dehydrate the toilet contents. These toilets are often referred to as "urine diverting dry toilets" or UDDTs. The conditions needed for pathogen or parasite elimination are not likely to be achieved in urine diverting dry toilets.

Another type of dry toilet is a "compost toilet." It isn't called a "compos*ting*" toilet, because that implies that composting is taking place inside the toilet. Since the toilets themselves don't create compost but only collect organic material, there's no point in calling them *composting* toilets. A *compost* toilet, on the other hand, is any toilet that collects toilet material so that it can be composted in a separate area apart from the toilet. And composting, as stated, requires human management, aerobic conditions, and the generation of biological heat. Since compost toilets collect material to be composted rather than dehydrated, urine separation is neither necessary nor recommended. Urine is a beneficial additive in compost piles and is easily and conveniently collected inside a compost toilet receptacle. There is no need to separate the liquids from the solids as in a UDDT. There are scores of dry toilets available on the market today, worldwide.

Continued on page 18

Finished compost from toilets at the Petion-ville Children's Academy and Learning Center in Haiti, is hygienically safe, odorless, and fly-free. The toilet building is shown on the opposite page. Rainwater is collected in a large black tank off the roof for washing purposes. The composting is conducted in bins immediately behind the building. Many toilets produce polluted water, waste, sewage, odors, maggots, and flies. Compost toilets produce compost, a valuable resource used to grow plants.

Urine diverting dry toilets (UDDTs) in Finland (opposite page) and in Mozambique, Africa.

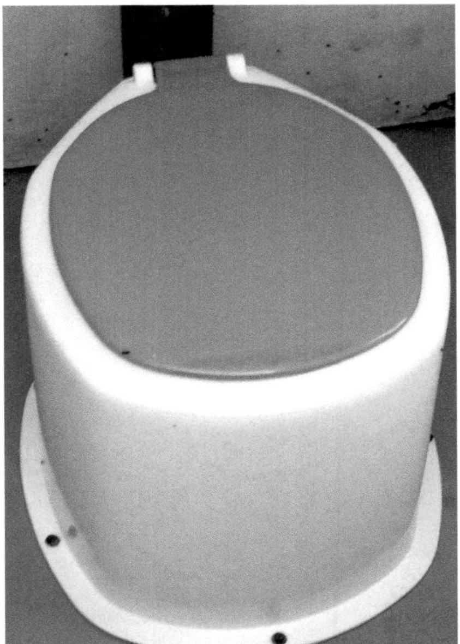

If the contents of the toilets are collected and then composted, they could correctly be referred to as compost toilets. Otherwise, they're dry toilets, not compost toilets. And they are certainly not *composting* toilets.

We humans are peculiar in that we are the only land animal that intentionally defecates in water. Water toilet users will seek out drinkable water to poop in even when there is little water to be found, such as during severe drought conditions. Water toilet cultures often don't have any viable alternative to defecating in water supplies, other than to revert to open defecation or pit latrines, which could be environmentally disastrous during a prolonged power outage, natural disaster, or other emergency.

A 5-gallon (20-liter) compost toilet receptacle and a bag of clean, finely ground, slightly moist organic cover material, such as sawdust, peat moss, sugarcane bagasse, rice hulls, or whatever you can grind up in your chipper/shredder, creates an instant compost toilet. If the contents of the toilet are regularly deposited into a compost bin, and a steady supply of cover material is available, such a compost toilet system can be used continuously for a lifetime, while producing no sewage. Compost bins can be built easily and quickly. A durable pallet bin can be built in 10 minutes. A wire bin or a straw bale bin can be erected in a very short time, too.

Drinkable water is arguably our most precious resource; without it we would soon die. Because very little water is used or required during the operation of compost toilets, human excrement is kept out of our water supplies and out of our environment. When we consider that 97 percent of the Earth's water is saltwater and that two-thirds of the fresh water is locked up in ice, we realize that less than one percent of the water on planet Earth is available as drinkable water. Considering that we have nearly eight billion people on Earth, and each person can produce about 25,000 pounds of excrement in his or her lifetime (11,340 kg, about a pound or 0.45 kg per day), according to Livescience.com), or, if you add it all up, about 640 billion pounds per year (290 billion kg, 2018 statistics), the idea that this giant mountain of excrement must be dumped into our drinkable water supplies on a continuous basis, as some sanitation professionals insist, is insane.

When we discard our excretions as waste, we create pollution and health hazards. When we compost them and return them to the soil, we create health benefits. Although the former situation is well known, most people don't know anything about the latter. One misconception is that fecal material, when composted, remains fecal material. It does not. When the composting process is finished, the end product is compost, not poop. When microbes eat your poop, they convert it into something else. Your poop is gone.

Human excrement is a substance that is widely denounced and has a long history of condemnation throughout the world. Ancestors who failed to responsibly recycle the substance created monumental public health problems. Consequently, the attitude that human excrement itself is dangerous has been widely held to the present day. This fear, fostered and spread by government authorities and others who know of no constructive alternatives to waste disposal, still maintains a firm grip on many people. A more constructive attitude is displayed by people with a broader knowledge of recycling manures for agricultural purposes. Informed persons know that the benefits of proper humanure recycling far outweigh any disadvantages from the health point of view.

Compost toilet users don't grow food with "human waste," nor do they grow gardens with "night soil." Instead, they feed microorganisms in a compost pile, not only with toilet material, but also with other animal manures, fruit peels, rinds and seeds, coffee grounds, meat, bones, fats, all sorts of food scraps, animal mortalities, garden residues, grass clippings, leaves, and so on. The microorganisms, over time, convert these organic materials into compost. People then add the compost to soil, which makes it available to plants. Then either they eat the plants or they feed the plants to animals; then they eat the animals or their by-products.

When compost toilet users relieve themselves, they aren't "taking a shit." They're giving a shit. Think about that for a moment.

Don't Say "Bucket"

There are two words that should never be used in association with compost toilets. One is "waste." A compost toilet is a device that recycles organic material; it does not collect or dispose of waste. It collects humanure, and it completely recycles it. No waste goes in, and none comes out.

The other word is "bucket." Some compost toilets utilize 5-gallon buckets as toilet receptacles. Others use drums, urns, barrels, bins, or any toilet receptacle that is watertight and manageable.

Five gallons, or approximately 20 liters, is a convenient capacity for easy handling by one person, and a 5-gallon container will hold approx-

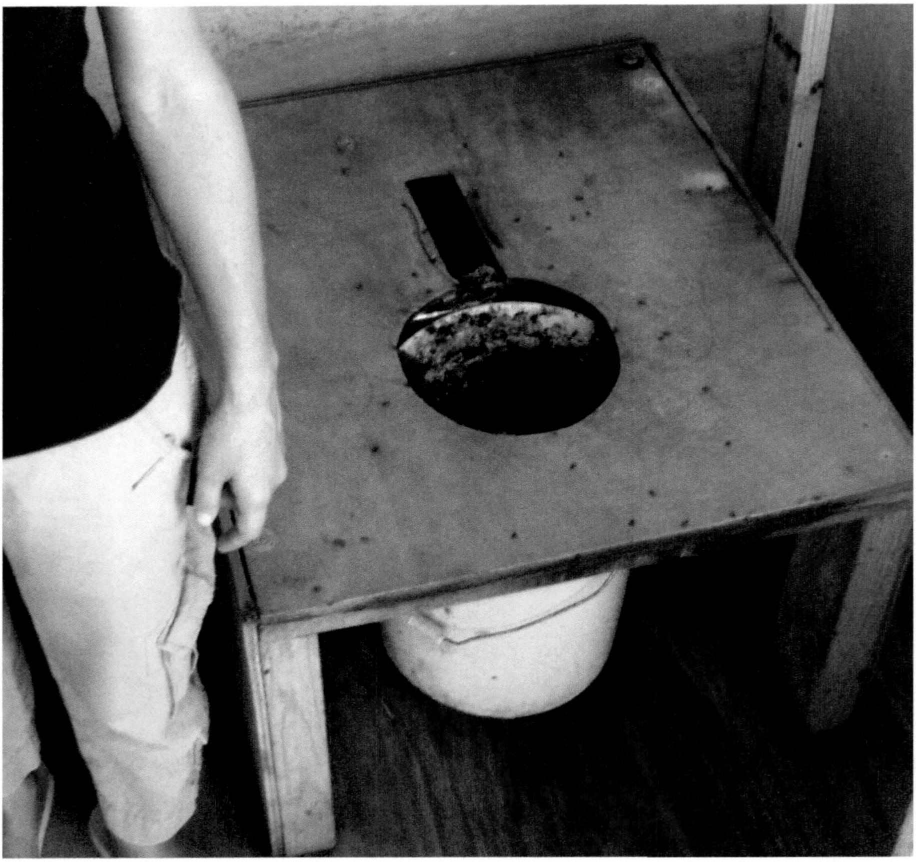

This "bucket toilet" in Haiti is an open, fly-infested, bucket of excrement that stinks like hell. Bucket toilets can easily be converted into odorless, fly-free compost toilets by the simple addition of a carbon-based cover material and a compost bin.

imately one week's excretions of one typical adult, assuming an appropriate cover material is used in the toilet. Five-gallon plastic buckets are easy to come by in some countries, such as in the US, where they can be acquired cheaply or, when recycled, for free. In other countries they are nearly impossible to find.

"Bucket toilets" however, are not compost toilets. Bucket toilets were commonly used in, for example, prisons, where inmates had to defecate and urinate in open buckets. No cover material was used, and the bucket contents were simply dumped outside. The bucket toilets smelled horrible, attracted flies, polluted the environment, and severely undermined the quality of life. They are still used in some parts of the world.

Bucket toilets date back generations, are widely condemned by sanitation professionals, and are not to be confused with compost toilets. Even when a compost toilet utilizes buckets as toilet receptacles, it is still not a bucket toilet; it's a compost toilet. It's a good practice to avoid the use of the word "bucket" altogether when discussing compost toilets. Compost toilets have receptacles or containers in which the toilet material is collected. When someone uses a compost toilet that has a 5-gallon receptacle, they are not "shitting in a bucket." They're using a compost toilet. When people use a flush toilet are they "shitting in drinking water"? Well, yes, they are, but we don't use that terminology; we say they're using a toilet.

Most people know little about compost toilets, but bucket toilets have a long history. They are not the same thing and are not to be confused. Choose your terms carefully when referring to compost toilets so as not to spread misunderstanding.

Human Nutrient Cycle

The "human nutrient cycle" is when a human consumes food (nutrients), then the human's *excretions* become food for other organisms. Those organisms convert the excretions into soil from which more food is grown. Humans eat *that* food, excrete, recycle, grow more food, and continue a natural, endless cycle. When we dispose of our excretions as waste, the cycle is broken. When we *compost* our excretions, we provide nutrients

HUMAN NUTRIENT CYCLE

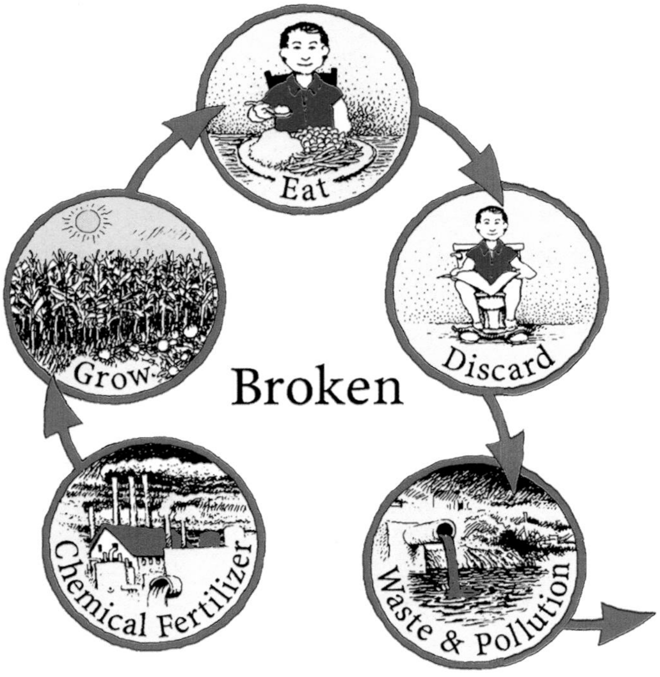

for beneficial microorganisms, which in turn convert them into compost, thereby returning the nutrients back to the earth where they provide soil fertility and crops, and the cycle continues unbroken.

The recycling of organic material should be taught to our children in schools and to adults in universities. Learning how to live on our planet in ways that are benign, sustainable, and symbiotic is critical to the long-term survival of the human species. What about fifty generations from now? If considering the future when conducting our daily affairs seems unrealistic, at least consider enough of the future to include your grand-children and great-grandchildren. The way we live today may have a significant impact on the way our descendants live in the future.

Government authorities may be unwilling to teach you how to make compost from toilet material. They often have no training in composting, nor do they understand how it works. Many government authorities in water toilet cultures have a nineteenth century attitude about human excrement, believing it is to be a dangerous waste that must be disposed of.

The world's billionaires, who have hoarded the Earth's wealth for themselves, are unlikely to be helpful. So far, the efforts to "reinvent the toilet" have been efforts to create a product that can be patented for profit. Many people who don't have toilets don't have money either. If reinventing the toilet is based on profit motives, it is unlikely to succeed in a manner that benefits the billions who need effective, affordable sanitation.

Organics recycling can be a profitable business when responsibly managed, on both large and small scales. Compost toilet construction, compost bin construction, and compost toilet training can also be profitable ventures for small local business enterprises. People without toilets can also construct compost toilet systems on their own, or maybe with a little help from their friends.

Above: This is a toilet at the Mekelle University main campus, Mekelle, Tigray, Ethiopia, where piles of feces littered the entryway to the toilet stalls. The pit latrines behind the doors had filled up, making the entryway the most attractive place to find relief. The photos above show the same hallway before and after compost toilets were installed. **Below:** The photos below show the same university toilet stall before and after it was converted into a composting system. It's amazing how little knowledge exists throughout the world about using composting as a sanitation alternative. These photos illustrate the stark contrast between an outdated, obsolete pit latrine sanitation system and a modern, biological, compost toilet sanitation system.

Photos provided by Samuel Autran Dourado.

Compost bins at the WeltHungerHilfe headquarters in Moroto, Karamoja, Uganda (above), contain months of collected toilet material without odors or flies. Microbes inside the bins convert the organic material into compost.

This is a compost toilet building at the Ayder Referral Hospital & medical school, Mekelle, Tigray, Ethiopia.

Photos by Samuel Autran Dourado.

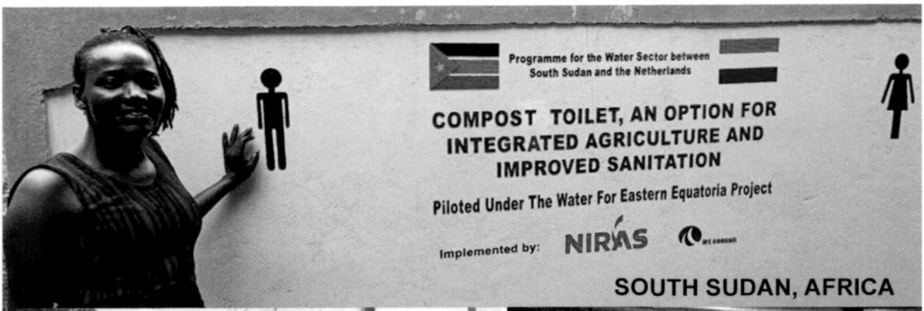

SOUTH SUDAN, AFRICA

KAMPALA, UGANDA, AFRICA

CHILD AND ADULT COMPOST TOILETS
Wayuu tribe, La Guajira, Riohacha, Colombia, South America. A WaterAid America project.

Photos by Samuel Autran Dourado.

Chapter Two

COMPOST TOILETS

Four things are needed for a compost toilet to function correctly. Without all four of these requirements, a compost toilet will not work:

(1) The toilet
(2) The compost bins
(3) The cover material
(4) Human management

Toilet material nourishes a compost pile. It provides much needed nitrogen and moisture. People with compost piles consisting primarily of leaves and garden weeds may not be observing heat being generated within the pile. Add nitrogen and moisture on a regular basis, and you will likely see the piles get hot.

Toilet material smells bad. Mother Nature makes things smell bad so we will bury them. Covering with dirt has been the age-old solution to bad odors. Decaying corpses smell horrible, but not if they're buried under

Compost toilets are introduced at the Kwale's girl's high school, Kenya, Africa. The toilets utilize 5-gallon (20-liter) plastic receptacles encased in a wooden cabinet. The hinged top allows for the easy removal of the toilet contents. Compost toilets such as these can be located wherever a convenient toilet is needed. This is an Aqua-Aero WaterSystems project. Photo by Samuel Autran Dourado.

earth. However, it's not only earth that blocks the odors of rotting organic material; ground-up plant material such as sawdust, sugarcane bagasse, and rice hulls work just as well. The difference is that ground plant material also sets the stage for composting. The microbes need the carbon in the plant material to balance the nitrogen in the toilet material. Soil doesn't provide the carbon, but plant cellulose does.

A compost toilet is used to collect the toilet material, and it's the simplest part of the system. The collection receptacles must be sturdy, waterproof, and durable. If the receptacles are being emptied by hand, they must be small enough to be handled by one or two people. One person can handle 5 gallons (20 liters) of capacity, two people can handle 15 gallons (60 liters) or more. If you're alone and 20 liters is too heavy, empty the receptacle when it's partially full; don't wait until it's too heavy to handle.

All urine, fecal material, and toilet paper go into a compost toilet, as does anything else that would normally go into a flush toilet. A compost toilet, although a type of "dry toilet," is not designed or intended to dehydrate the toilet material, as are most commercial dry toilets. Compost toilets collect toilet material for composting; therefore, all urine is also collected. You can also throw in the cardboard tubes from the center of toilet paper rolls. You can vomit in the toilet. Just don't put food scraps in the toilet because you can risk a fruit fly infestation. Put food scraps in the toilet *after* the toilet receptacle has been removed from the toilet cabinet, and a lid has been placed on it. It's all going to the same compost pile anyway.

You can construct your own toilet cabinet for very little money out of scrap wood. You can often find recycled plastic containers, such as 5-gallon (20 liter) buckets and 15-gallon (60 liter) drums, available cheap or free. You should construct the toilet to fit the receptacles, so make sure you have several receptacles that are the same size; otherwise they may not all fit.

The author uses five 5-gallon receptacles in his bathroom, one in the toilet, and four on standby with lids. When four are filled and set aside, with lids attached, they all go to the compost bin at the same time while the fifth remains in use. Although they're "full," there is still about a gallon (4 liters) of room in each receptacle to add food scraps from the kitchen compost pail *after* the receptacle has been removed from the toilet and a

Continued on page 33

This is one of the author's "Loveable Loo" style compost toilets located in an office bathroom. The metal container holds sawdust from local sawmills, which is used as the cover material. The lid on the cover material container keeps the sawdust from drying out and prevents cats from using the container as a litter box. The small broom comes in handy for whenever sawdust gets spilled or messy. The toilet receptacle consists of a 5-gallon (20 liter) plastic container that lifts out the top of the toilet cabinet. When a receptacle is filled, it is removed, a lid is placed on it, and then it's set aside to be taken to a compost bin when convenient. This office building has two such toilets. The contents of the receptacles are fed into a nearby compost bin about every six weeks. This usually involves feeding the contents of 10 to 15 receptacles to the compost pile all at once, requiring about a half hour of time from one person.

BUILD A SIMPLE COMPOST TOILET

1" = 2.54 cm

1. Start with 4 identical receptacles and a standard toilet seat.

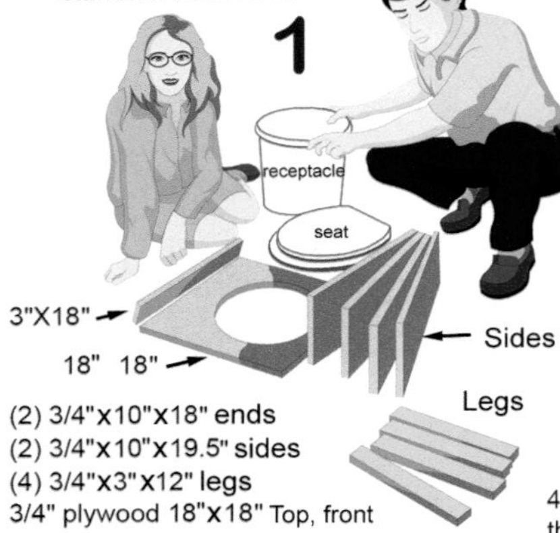

(2) 3/4"x10"x18" ends
(2) 3/4"x10"x19.5" sides
(4) 3/4"x3"x12" legs
3/4" plywood 18"x18" Top, front
3/4" plywood 3"x18" Top, back
(2) Hinges 3" (7.6 cm) or similar

2. Screw the four boards together.

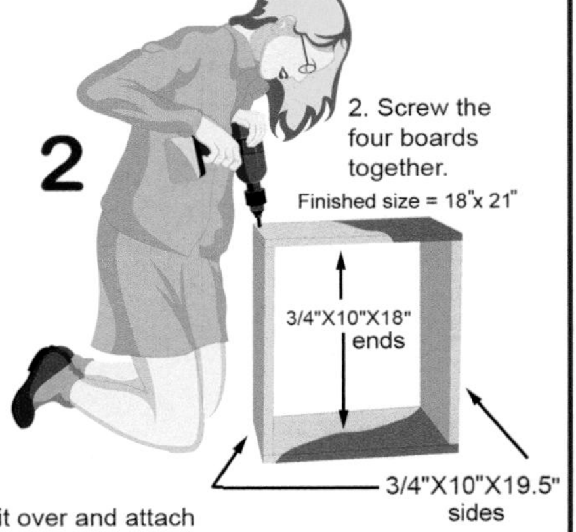

3. Screw the 3" X 18" board to the box. Leave the 18" X 18" plywood top attached only by the hinges.

4. Flip it over and attach the four legs to the inside of the box. The receptacle must protrude through the plywood top by about 1/2". Adjust legs accordingly.

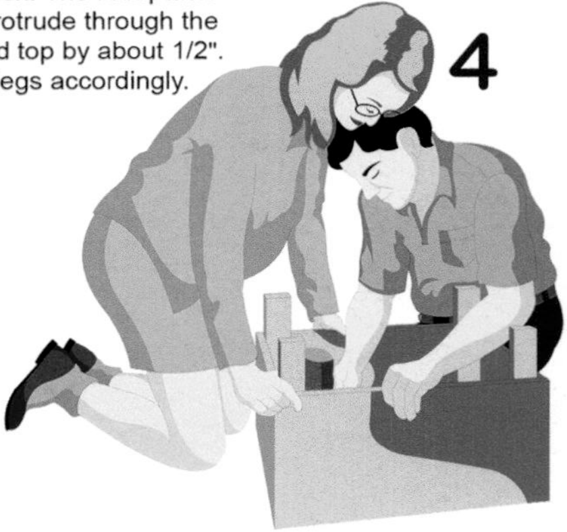

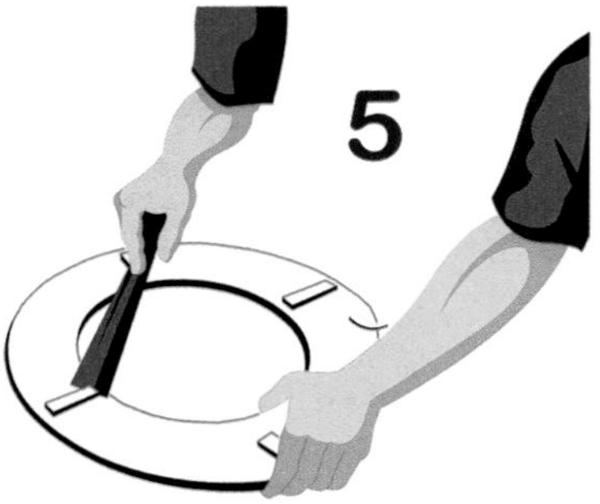

5. Swivel the plastic bumpers sideways so the top of the receptacle rim will almost touch the toilet seat ring. The adjusted toilet seat is shown at right. You will have to drill a new hole for each bumper.

Black dots show original holes. Red dots show where the new holes were drilled underneath the bumpers. The holes accommodate the pegs on the underside of the bumpers.

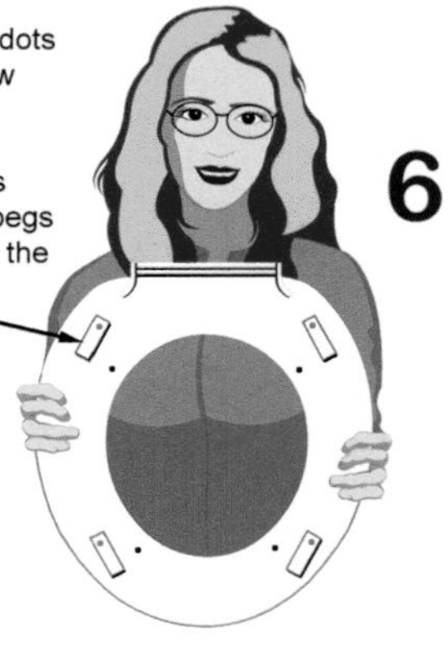

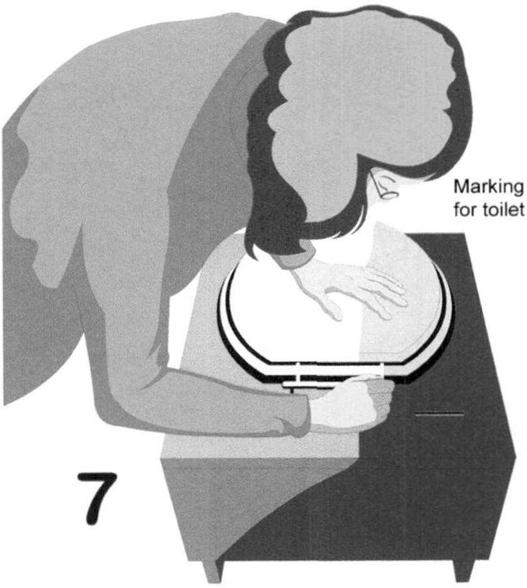

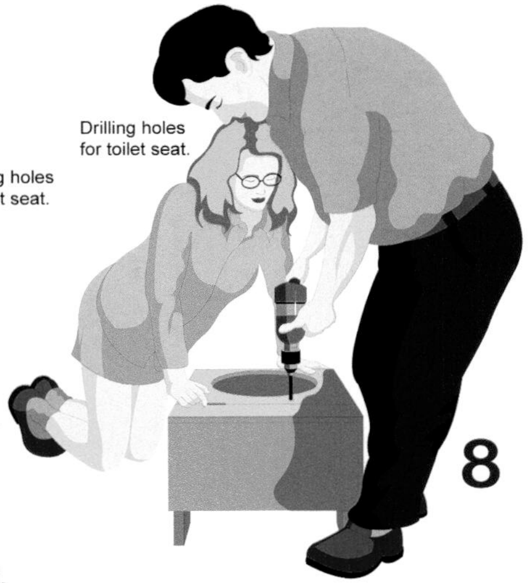

Mark holes for the toilet seat attachment. A hinged compost toilet cabinet will be 18" wide by 21" long.

Cut the hole in the 18"x18" plywood so it allows the top of your toilet receptacle to fit through it. Set the hole back from the front of the plywood 1.5 inches. These two details are important! Your receptacle must protrude through the plywood top by about 1/2 inch. You adjust this height when you attach the legs. The toilet is built to fit your receptacles, so make sure you have plenty of same-size receptacles.

When screwing the legs to the inside of the box, adjust the height so the receptacle will protrude through the top of the cabinet about 1/2 inch. This allows for a close fit against the toilet seat so all of the toilet material goes down into the toilet and not over the top of the rim. This is why the bumpers were swiveled sideways (to make room for the receptacle rim).

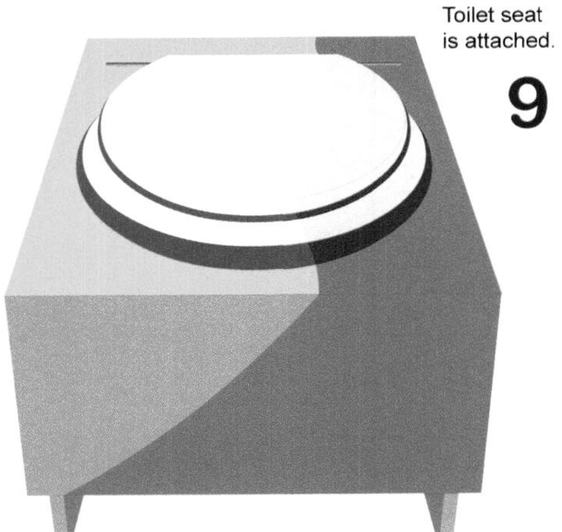

9. Attach your toilet seat using the two bolts that came with the seat. Stain, varnish, or paint the wood. You now have a compost toilet!

GiveLove

snug lid attached. Topping off with food scraps will make the receptacle heavier, but if that's not a problem, this is a convenient way to get all your organic material out to the compost bin at the same time.

Why not just put the compost toilet over top of the compost pile? Then nobody would have to dump any toilet receptacles! Because most people like their toilet indoors, where it's convenient, secure, and comfortable, year-round. Compost *piles* can only be located in limited locations, but compost *toilets* can be located anywhere a flush toilet can be located. In fact, compost toilets are much more versatile than flush toilets, as they can be located where there is no water or electricity; they don't need to be hooked up to anything, no drains or vents, for example, and they can be constructed and operated for very little money. They are also odor-free when properly managed, so they can even be located immediately next to a bed, which can greatly benefit the elderly or bedridden.

In the author's own home, he has a compost toilet in an upstairs office, another in an upstairs guest bedroom, and another in a downstairs bathroom. He also has one in a separate guest quarters and two in his business office. All are odor-free. Why would anyone want to have to go outside and climb up on top of a compost pile, or go to a single indoor location to use a toilet when the toilet can be located indoors in any convenient place?

All one has to do once a week (for a family of four) or once a month (for a single person) is deposit the contents of four or five 20-liter compost receptacles into an outdoor compost bin, a half hour task. If a compost bin is located underneath a house, it will still need to be emptied, you're still going to produce the same amount of compost, and you'll eventually want to get it to your garden, flower beds, or fruit trees. It may not be convenient to manage the compost when it's under a house, or to have to move it up to ground level during the growing season if it's in a basement. The more difficult we make the process, the more likely it will fail.

Continued on page 38.

Opposite page: Compost toilets are being designed for the NGO Kolkata Priyodorshini by Samuel Autran Dourado for community use in Kolkata, India in 2018. First, receptacles are sourced, then the cabinets are built to fit the receptacles, which, in this case slide in and out of the front of the toilet cabinet. Note the color-coded nature of the containers in the bottom photo. Light blue containers are for cover material, dark blue containers are toilet receptacles.

A compost toilet made from a repurposed plastic drum in Lira, Uganda, Africa (a GiveLove.org project).
Photos by Samuel Autran Dourado.

A resourceful young man in Mozambique repurposed a plastic chute from a commercial dry toilet and converted it into a compost toilet (above). Another type of dry toilet that has been converted into a compost toilet is shown below.

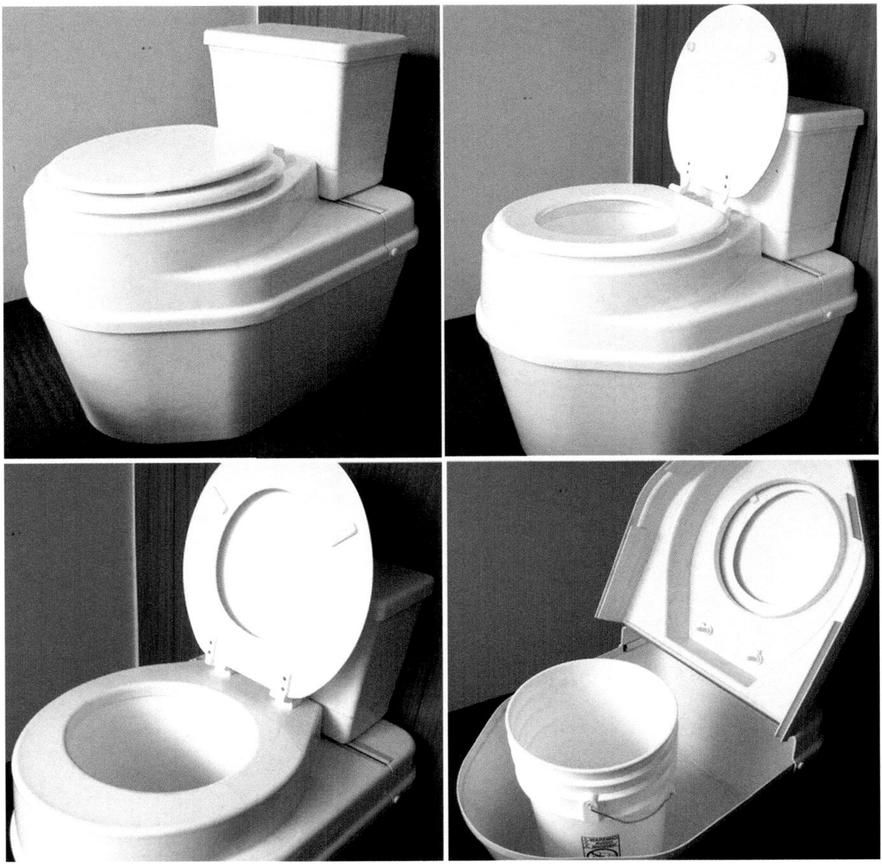

Above: Compost toilet built for the Wayuu tribe in La Guajira, Riohacha, Colombia, South America, a project with WaterAid America in 2017. **Opposite:** Compost toilet building under construction, also in Colombia. Photos by Samuel Autran Dourado.

How do we eliminate odors? It's easy: always keep the toilet contents covered. If you're using an adequate amount of *appropriate* cover material, your compost toilet will be odor-free, as will your compost bins. If your nose picks up bad odors, or your eyes see flies, you're not keeping the compost covered correctly. You need a nose that works and eyes that work to be able to manage a compost toilet system. A brain that works helps, too. Compost toilets are "the thinking person's toilet."

Add more cover material if you smell anything. Make sure it's fine enough, and has some residual moisture content, when covering your toilet contents. If it's light and fluffy and dry like bagged wood shavings, it will not be as effective an odor blocker in your toilet, and you will need to use more of it, filling your toilet too quickly. Wet it down, and let it rot, even for years if needed; *then* use it for covering the toilet contents.

Always cover the contents of the *compost bins* thoroughly as well. If you smell odors, add more cover material. The cover material can be deep; it won't hurt the compost. You're going to move it aside as you add new or-

Locally abundant rice hulls are used in Tipitapa, Managua, Nicaragua by the NGO Sweet Progress as cover material in the school and village compost toilets.

ganic material anyway. You should never see anything on top of your compost bin but the cover material.

If your toilet or compost bin smells bad or attracts flies, add cover material until it doesn't. Still smell something? Add more cover material. When you add enough cover material to block the odor from poop in a toilet, you will have effectively thrown off the balance between the nitrogen from the toilet material and the carbon from the cover material. There will be too much carbon. This is why the urine must also be collected in the toilet. Although a toilet receptacle may look "full" of poop covered in carbon material, it will still accept a lot of urine. The urine is what adds the extra nitrogen and moisture to balance the carbon cover material. As an experiment, fill a toilet receptacle with cover material, then pour water into it until it's filled. Measure how much you added. You may be surprised at how much liquid (i.e., urine) can still be added to what looks like a full toilet receptacle!

With enough toilet receptacles, a compost toilet system can be used for

A community worker helps establish compost toilets at the State Ministry of Physical Infrastructure in Torit, South Sudan, Africa (a project of NIRAS development consulting). Photo by Samuel Autran Dourado.

Gardens benefiting from humanure enriched compost in Africa, at Five Star Academy in Nairobi, Kenya (top, a project of Humanure Kenya in collaboration with GiveLove.org) and in Namalu, Uganda, bottom. Bottom photo by Samuel Autran Dourado.

any number of people. If you're using one in your home and you're suddenly visited by many people, you'll be happy to have empty receptacles on hand to replace any that may fill up. You will not have to empty any toilet receptacles until after your company leaves, because you can set them out of the way, with lids, and then empty them when it's convenient.

This is the same procedure used for large gatherings, even tens of thousands of people. Larger compost toilet receptacles are utilized, perhaps 50 gallons (about 200 liters) or more, perhaps handled by machinery, and as they fill, they're moved out of the toilets and replaced with empty ones. Lids are placed on the full containers when they're set aside and waiting to be emptied into a compost bin. The contents of the full containers are deposited into the bins when convenient, which could be days, weeks, even months later. A designated person, or persons, remains responsible for monitoring and maintaining the toilet system.

When managing compost toilets for groups, always be prepared for the unexpected and have extra toilet receptacles and cover material available. For every full compost container removed from a toilet room, a full, same-size container of cover material will likely need to be provided.

Most of the toilets shown in this chapter have been ones with small capacities. This is because they are managed by individuals who must carry the toilet receptacles from one place to another. Also, smaller toilet receptacles allow for easier covering of the contents. However, larger toilet receptacles utilized when providing sanitation for larger groups of people work according to the same principles. The contents are covered with a carbon-based plant material and all urine, fecal material, and toilet paper are collected in the same container. When the containers are full, they are set aside to be composted at a convenient time. Machinery may be required to move and handle the heavy toilet receptacles. Compost bins must be larger, and compost moving machinery may also be required.

No matter how many people are being serviced, where there is a will, there is a way.

Compost bins are woven from plant material in Torit, South Sudan, Africa (a project of NIRAS development consulting).
Photo by Samuel Autran Dourado.

Chapter Three

COMPOST BINS

Compost bins are the second necessary element of a compost toilet system. You will need at least two bins, because once you fill one, you will have to allow it to "age" or "cure" for a period of time, approximately one year. In the meantime, you will need a second bin to fill.

The bin holds the organic material vertically aboveground so that dogs, goats, horses, and other animals can't get into it. Holding it aboveground helps maintain an aerobic system. A bin also provides side walls, thereby preventing exposed compost surfaces.

In many cases you can simply size the bins so it takes about a year to fill them. If you're composting for a larger group, you may need to use several bins or larger bins. In some situations, you will be able to fill large bins relatively quickly. In all cases, once the bin is full, even if it fills in one day, allow it to age for approximately one year, give or take a couple of months. Do not try to rush the curing phase.

A standard family bin size is roughly 5 feet square and 4 feet high (1.5 meters square by 1.2 meters high), or roughly 2.8 cubic meters in volume. Wood pallet bins can be smaller. Four average recycled US pallets, standing on edge, will make a bin approximately 4 feet by 4 feet by 3 feet high (1.2m × 1.2m × 0.9m), or 1.3 cubic meters.

You can construct a quick four-pallet bin in ten minutes using wood pallets standing on edge. Just lean them against each other and run a couple of screws in each side to hold them together. If you don't have screws, tie them together with something. If you need larger bins, such as for a school, make them two pallets wide, but not much wider. You need to be able to reach into the middle of the bins from either side with a shovel or tool to manage the compost. You can make them as long as you want, however.

Compost bins can easily be built to be animal proof. If small animals such as rats are a problem, the compost bin can be lined with wire mesh on all sides, underneath, and on top. The bins can have side walls such as wire, blocks, bricks, straw bales, wood boards, or similar barriers to keep

Newly constructed compost bins in the Zombo district, northern Uganda, a project of International Medical Outreach, are shown above and below. The bin above is immediately adjacent to the toilet building, making it very convenient to compost the toilet material. When the bin is near the toilets, transport labor is minimized.

Photos by Samuel Autran Dourado.

out larger animals. A simple piece of wire fencing cut to fit the exposed top of an active compost pile will keep animals from digging into it while still allowing rainwater to keep the pile moist. Thorny branches will discourage horses from eating the grass that may be covering a compost pile. With adequate cover material on the pile, there will be no flies or odor.

Position the bins on soil; a concrete base is not needed. The soil/compost interface is useful for several reasons. It provides a biological conduit for micro- and macroorganisms to enter and exit the compost pile. It also provides an area for compost microbes to reside after the bin is emptied; these microbes help inoculate the next pile. The upper few inches of soil also act as a buffer for excess liquid, absorbing it if needed. Moisture contained in the upper few inches of topsoil located underneath a compost pile is not a public hazard. Although fecal contamination of drinking water supplies has been a worldwide problem for ages, this is caused by raw sewage seeping into water wells, streams, rivers, and lakes from pit latrines, cesspools, septic tank leachate, sewage overflows, and open defecation. Septic tank leachate is notorious for contaminating groundwater, and even public sewage systems are well-known sources of water pollution. A compost pile is none of these things and should not be treated as if it were.

Requiring concrete bases underneath compost piles greatly increases the cost of making the bin. If people needed a concrete pad under their piles, the cost could be prohibitive and discouraging, especially at village levels, causing the cost of a compost operation to be out of reach for people of meager means, the very people who need compost the most. Many people can't afford a concrete floor in their home. They shouldn't have to have a concrete floor under their compost pile. Many large-scale composting operations are done on bare soil, including sewage sludge composting.

Dish the soil base to create a shallow bowl underneath the compost pile. Take the dirt you dig out — and it doesn't need to be much — and throw it up against the *inside* of the bin walls around the bottom edges. You now have a depression underneath your bin providing extra insurance against any leachate seeping out the bottom. There is no reason to believe that some leachate collecting *underneath* a compost bin is in any way a cause for concern. It's absurd that some regulatory professionals may prohibit compost piles on soil bases, even when the compost toilets are replacing pit latrines and removing all of the pollution that would otherwise be collecting in holes in the ground. Composting can be a steep learning curve for sanitation and health professionals who are taught that human excrement is a dangerous waste material, and only a waste material, and must therefore be disposed of. The concept that toilet material is recyclable may not exist in their minds. This can, however, change through time, education, and experience. If you know an authority who needs this book, the author will provide a free copy, no questions asked. Just email the publisher the mailing address (US only) for a paper copy, or the email address for foreign residents, who will be provided a digital download.

Before adding any toilet material to your bin, first create a "biological sponge" in the bottom. This is a cushion of grasses, weeds, leaves, hay, straw, or whatever else you have on hand, or whatever you're using for

Education is key to successful community compost-based sanitation. Most people know nothing about this process, but the social stigmas of dealing with human excretions are soon overcome when the process is seen and understood. This is in the Zombo district of the northern region of Uganda (an International Medical Outreach project).
Photo by Samuel Autran Dourado.

cover material, at least 18 inches (1/2 meter) thick, or more. Thicker is fine; it will compress down and disappear in the finished compost.

Put enough of a biological sponge in the bottom so that you can create a depression in it to place your first organic deposit, then rake the material back over the deposit and add more cover material. The fresh deposit, which may be toilet material, a dead animal, food discards, or a combination thereof, is now buried in the cover material. When adding additional material to your bin, use a dedicated tool such as a shovel, fork, or rake; peel open the cover material; dig into the existing compost to again create a depression; dump the fresh material into the depression; then rake the compost and the cover material back over it. Add more cover material. Keep your compost pile flat on top. This prevents rain runoff and makes it easier to keep the pile covered.

Rinse or wash your toilet containers and deposit the dirty water into the bins. Keep a compost thermometer in the center of your *active* pile, the one you are currently adding material to, if possible. Twenty-inch "backyard" compost thermometers are inexpensive, and they give you a constant reading letting you know what's going on inside your pile. More expensive, longer, professional compost thermometers are available and recommended for researchers, compost managers, and instructors. Pull the thermometer out before adding new material.

By adding material into the center of the compost pile you inject the new material into the most active part of the pile. You also cover it thoroughly, not only with cover material, but also with existing compost and, when raking the cover material aside to make room for new organic material, you're creating a cover material envelope around the outer edges of the compost, thereby wrapping the compost in cover material, like a blanket. This insulates the pile, keeps the outside edges of the compost from cooling down prematurely, and keeps compost from falling out of gaps in the walls, such as can happen with pallet bins. By using a cover material envelope around your compost, you can use any kind of open-bottomed bin: wood, block, brick, metal, or plastic, and holes, spaces, or gaps are *not* required in the side walls for aeration. Air is entrapped in the cushion, as well as in the organic material itself.

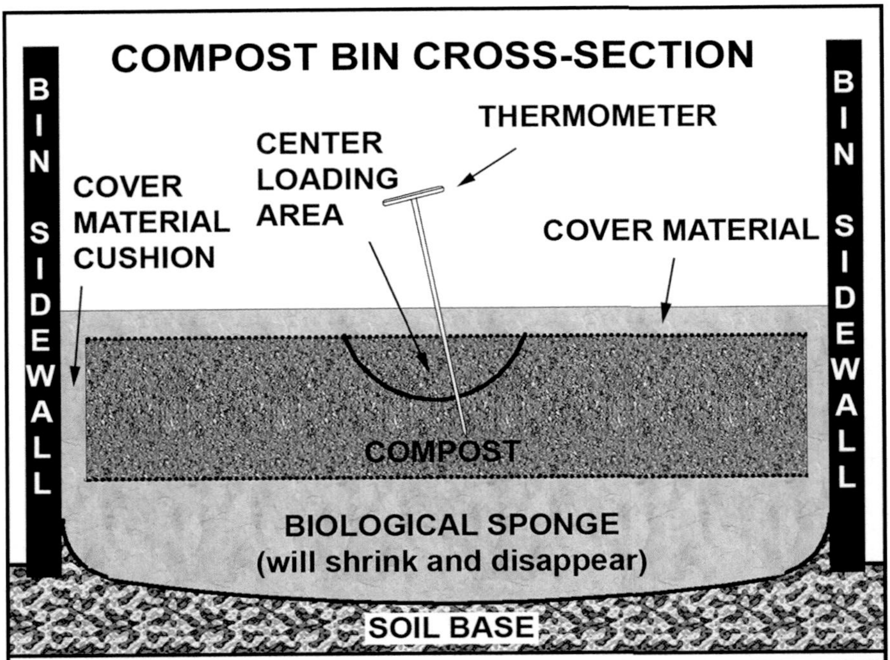
COMPOST BIN CROSS-SECTION
BIN SIDEWALL
BIN SIDEWALL
THERMOMETER
CENTER LOADING AREA
COVER MATERIAL CUSHION
COVER MATERIAL
COMPOST
BIOLOGICAL SPONGE
(will shrink and disappear)
SOIL BASE

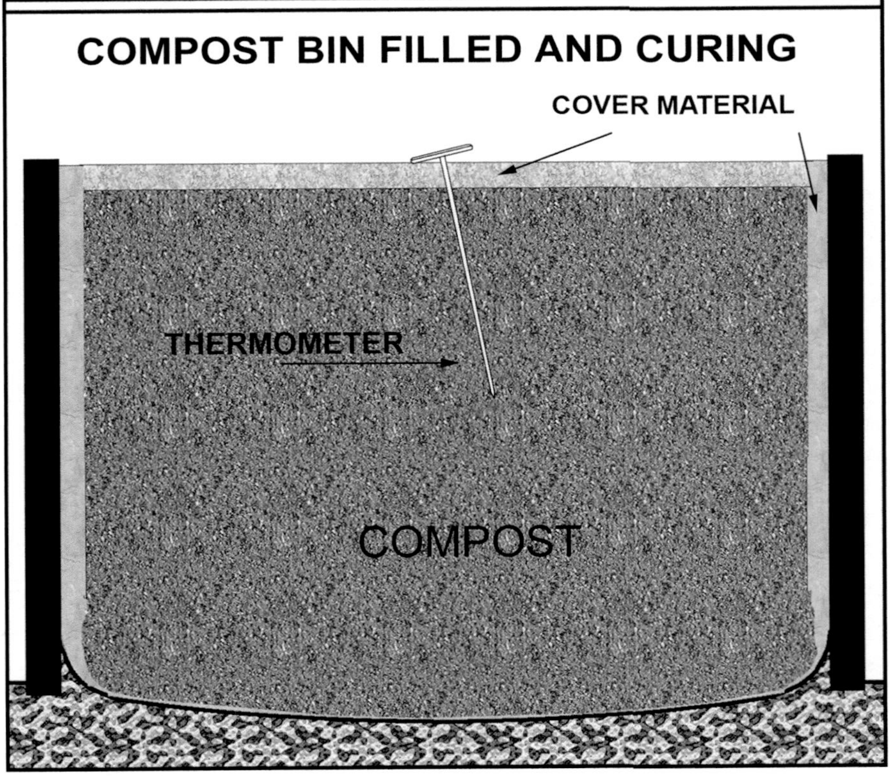
COMPOST BIN FILLED AND CURING
COVER MATERIAL
THERMOMETER
COMPOST

You could call this "center feeding" a compost pile, as opposed to "layering" that seems popular in some composting circles. Layering is loosely defined as adding new material to the top of the compost pile in flat layers and is not recommended when composting toilet material.

If you have rectangular straw bales, you can peel off "chips" or "flakes" from the ends of the bales, which are flat sections a few inches thick, and position them against the inside of the bins to line the interior walls before adding your organic material. This insulates your pile, keeps compost from falling out between gaps in the sidewalls, and provides an aerated space enveloping the compost.

For a household, a three-bin compost system is good because the center bin can be constructed with a roof or covering for storing cover materials. By keeping the cover material (straw bales for example) dry, it won't freeze and will remain available for use even during frozen winter months. When you have a sudden large quantity of cover material available, such as a seasonal influx of grass clippings, weeds, or leaves, you can also place them in the center bin for storage and use them to cover the compost as needed. If

Continued on page 53

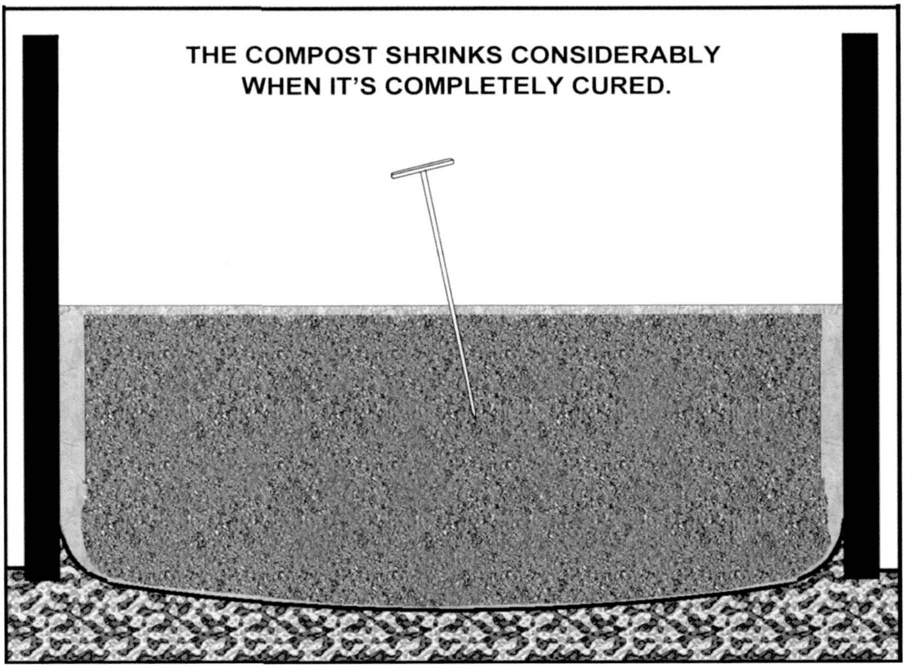
THE COMPOST SHRINKS CONSIDERABLY
WHEN IT'S COMPLETELY CURED.

HOW TO CONSTRUCT A

3-SECTION COMPOST BIN

1 inch = 2.54 cm

1. Dig eight holes about 24 inches deep and drop in 4x4 posts. Backfill with soil (and concrete mix, if available). Posts are about 5 feet (1.5 meters) apart. Leave the four center posts full length. Cut the four outer posts to a height of about 4 feet (1.2 meters).

2. Vertically plumb and then brace the posts. Nail a 4-inch by 4-inch (10 cm by 10cm) header across the four high posts.

3. Screw or nail boards to the posts as shown.There is no need to leave a gap between the boards. If you're in a cold climate with frost heave, leave about 2 inches (5 cm) between the bottom board and the ground.

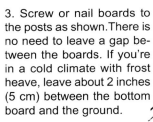

4. Cut the rafters and build a simple gable roof (or a simpler shed roof if you prefer, although the two front or back posts would need to be longer to allow for roof slope). The ground posts should be rot resistant. The sidewalls should not be treated with toxic chemicals. Periodically, the wall boards will need to be replaced. The central roof keeps the cover material dry so it won't freeze in a cold climate. The entire compost bin can be built from recycled lumber, if available.

3-SECTION COMPOST BIN (cont.)

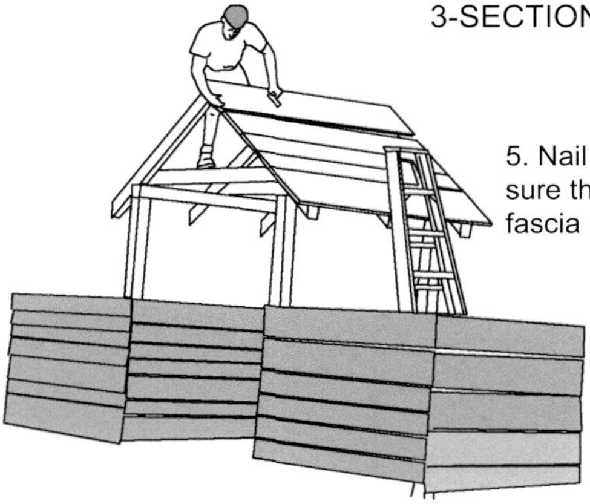

5. Nail roof boards to the rafters. Make sure the rafter ends have plumb cuts so a fascia board can be attached.

6. Install fascia boards, then install the finished roofing. Recycled slate is an excellent roofing material.

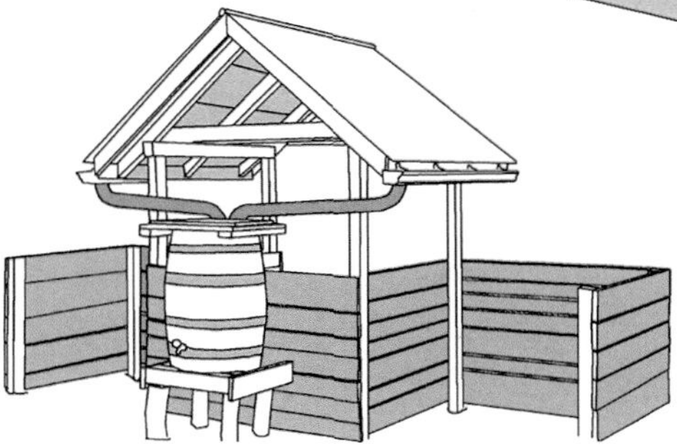

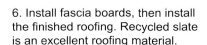

7. Install the rain spouting. Install a rain barrel. You will have to drain the barrel during freezing weather.

The author's three-chambered compost bin in Pennsylvania, USA, includes a center chamber for storing cover material, usually straw bales. The bin on the right is being added to (the "active" bin), while the bin on the left is beginning to be emptied out. Rainwater is collected off the center bin roof and used to wash toilet receptacles.

VARIOUS
COMPOST
BINS

you use poisonous chemicals on your lawn grass, don't put your grass clippings in your compost.

When the compost pile is fully built, it is covered with a generous layer of straw, leaves, grass clippings or other clean material, without weed seeds, to insulate the pile and to act as a biofilter; then it is left to age, undisturbed. No turning is needed. Never add *anything* to a compost pile in this resting or "curing" stage! Just leave it alone.

The previous year's bin should have been emptied by the time the second bin has been filled, and a new batch of compost can now be started in that bin, following the same procedure as the first — with a concave earth floor, biological sponge, cover material envelope, and center-feeding procedure. When that bin is nearly full (about a year later), the other one can begin to be emptied onto the garden, berries, orchard, or flower beds. If

Compost bins can be made from a variety of materials. What's important is that they vertically hold the organic material aboveground while preventing access by nuisance animals. These are in Torit, South Sudan, Africa (a NIRAS project).

Photo by Samuel Autran Dourado.

Above is a compost toilet stall used at a public festival in Pennsylvania, USA. The compost manager keeps a bin of sawdust (below) for use as cover material next to the toilet stall. The compost bins are located on either side of the sawdust bin. Compost toilets will not work without a carbon-based cover material to be used in both the toilets and in the compost bins. Compost toilet systems must be well-managed to provide a steady and reliable source of cover material.

you're not comfortable using your compost for gardening purposes, use it for flowers, trees, shrubs, or bushes.

A compost pile can accept a remarkable amount of organic material. Even though the pile may look like it's filling fast, it may soon shrink and leave room for more material. More than likely, the compost pile will keep taking more material as you add it because the pile is continuously shrinking. If for some reason your compost pile does suddenly fill up and you have nowhere to deposit the organic material, then you will simply have to start another compost bin. Four wooden pallets on edge, straw bales, bricks, blocks, or wire fencing will make quick bin walls in an emergency.

The system outlined above may not yield any compost for two years: one year to build the first pile and an additional year for it to age or cure. However, after the initial two-year start-up period, a family of four can expect to produce about a cubic meter of compost annually.

Everything is "homemade" in a Karamojong village in Moroto, Karamoja, Uganda where the compost toilets are located inside mud huts shown above at left, and the toilet material is composted in reed bins nearby, as shown at right (a WeltHungerHilfe project).

Toilet material from the Ayder Referral Hospital & medical school in Mekelle, Tigray, Ethiopia, is deposited into compost bins to feed microorganisms. Water used to wash the toilet receptacles is also deposited into the bins. Unsanitary wash water should never be discarded into the environment. Use the same water to wash multiple receptacles, then discard it onto the compost pile. Photos by Samuel Autran Dourado.

If you're composting toilet material from a population with endemic diseases, such as, for example, hospital residents, an additional year-long curing period should be considered, if room for additional compost bins is available. In that situation, after a bin is filled, it is left to rest for two years. This system will create a longer lag time before compost is initially available. If in doubt about the hygienic safety of any compost, either test it for pathogens in a laboratory, or use it agriculturally where it will not contact food crops, and wear gloves when handling it.

Finally, all organic material goes into the same bin! Food scraps; toilet paper; toilet material including urine; animal mortalities; stale beer; discarded meat, bones, and fat; citrus peels; rinds, husks, and pits from fruits; the list is endless. If you have animals and want to add the animal manures, that's fine, although you may need additional bins because they may fill up faster, depending on the size and quantity of the animals you have.

A wire compost bin at a children's school in Moroto, Karamoja, Uganda has been primed with a "biological sponge" in preparation for deposits of toilet material from the school's compost toilets, which are located in a small building nearby. This is a WeltHungerHilfe project. Photo by Samuel Autran Dourado.

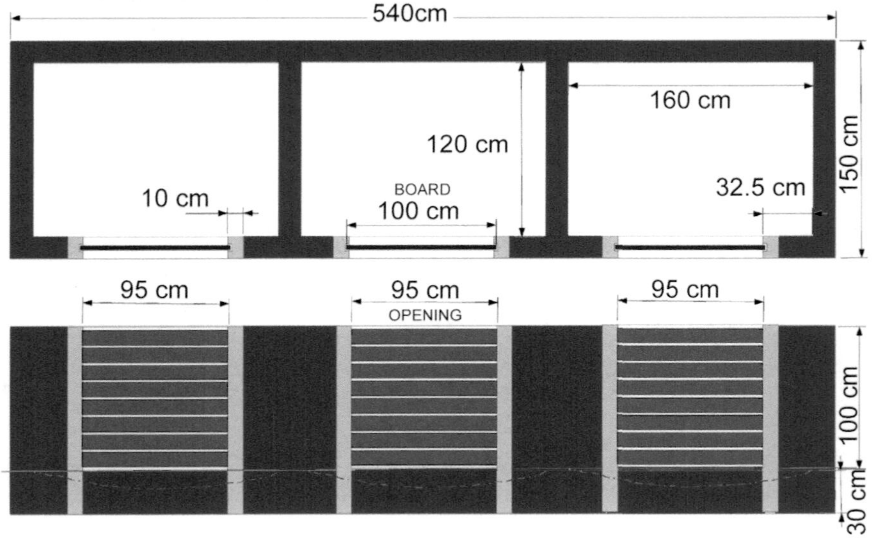
540cm
160 cm
120 cm
150 cm
10 cm
BOARD
100 cm
32.5 cm
95 cm
95 cm
OPENING
95 cm
100 cm
30 cm

Above and below: Compost bins can take many forms and can be expanded as needed when composting for larger populations. The bins above are at a the Kwale girl's high school, in Kwale, Kenya, Africa (an Aqua-Aero WaterSystems project), while the bins below are at the Ayder Referral Hospital & medical school in Mekelle, Tigray, Ethiopia. **Opposite**: Brick bins made for a compost toilet at the State Ministry of Physical Infrastructure in Torit, South Sudan, Africa are termite proof and will last a very long time. This is a NIRAS project. Photos and drawing provided by Samuel Autran Dourado.

Leachate

Remember that microbes don't walk, they swim. Keep your piles moist. Compost requires a lot of moisture and prefers to be damp. Piles that are dry become inactive and cannot heat up.

Evaporated moisture is one of the main reasons that compost shrinks so much. Compost piles are not inclined to drain moisture unless subjected to an excessive amount of rain or other influx of liquid. Most rainwater is absorbed by the compost, but in areas of heavy rainfall a roof or cover can be placed over the compost pile at appropriate times to prevent leaching.

A correctly managed compost bin will expose nothing but the cover material. At no time is the organic material inside the bin accessible to flies. Even if the bin has large gaps in the side walls, the cover material envelope that surrounds the compost contains the compost inside the bin, allowing nothing to escape.

This roof can be as simple as a piece of plastic or a tarp. You can also just add more cover material on top of the pile to protect it from heavy rains. The biological sponge acts as a leachate barrier too.

You can create leachate if you use too much wash water to clean toilet receptacles, because the wash water should be deposited into the pile. If your compost seems too wet, add more dry material into your compost mix when building your pile. And don't use too much wash water. If you're washing a dozen 5-gallon (20-liter) receptacles at once, for example, a gallon (4 liters) of water is enough to provide an initial rinse for all of them. Dump the gallon into the first receptacle, rinse it, dump the same water into the next receptacle, rinse, and repeat. Then dump that water into the compost bin. Use another gallon of water with a small amount of soap to wash all twelve 5-gallon receptacles. Dump that water into the compost bin, then use another gallon to give them all a final rinse to get the soap washed out. Dump that into the bin, too. Soap won't hurt your compost. You will have used 3 gallons (11.5 liters) of water to clean 60 gallons (227 liters) of toilet contents, which is about what one adult will produce in three months. A gallon of water per month per person (a liter per week) for toilet sanitation is not unreasonable. And all the soiled water gets processed through the compost pile and returned harmlessly to the earth.

Another way to process excess wash water is to fill a watertight container with dry cover material such as sawdust or bagasse, then pour the soiled wash water into the container. Even though the container is full of cover material, it will still take a lot of liquid. This is where bone-dry cover material is desirable. When the container fills up with the wash water, put a lid on it and let it sit for a while, maybe days, so the cover material can absorb the liquid. Eventually, dump the entire contents into the compost bin. If it's still too wet, drain it before dumping it. The drained water should go into another container of dry material, and so on.

The above-mentioned technique allows you to compost liquids of any type, including urine, and black water from latrines. Soak the liquid into sawdust or bagasse or something of the sort, drain it, then dump it into compost bins, ideally with other material such as food scraps.

If you don't like washing toilet receptacles and don't mind spending

Top: Compost bin at the Ayder Referral Hospital & medical school in Mekelle, Tigray, Ethiopia. (Photo by Samuel Autran Dourado) **Bottom:** The compost bin in Dongobesh, Tanzania, on the right, decided not to wait for the garden. The bin on the left is the active bin (the one being added to). Photo by Joseph Lawrence Lacha.

the money, buy compostable plastic bags for inside your toilet receptacles. That eliminates much of the wash water. Some people use newspapers to line their toilet receptacles. Then, when they empty the containers, the newspaper and the toilet material go into the compost pile together.

You may find that your compost can become too dry rather than too wet. You can add liquids to your compost to stimulate microbial growth. Especially effective are liquid food by-products such as discarded beer at breweries or the spent wine that remains after distilling wine when making brandy. Gray water from sink and bathtub drains, kitchen wash water, and other typically discarded household liquids may also rehydrate your compost piles when they dry out due to hot, dry weather. For compost organisms to be active enough to produce heat, the compost should have the moisture of a squeezed-out sponge.

Five Star Academy near Nairobi, Kenya, is a project of *Humanure Kenya*, where toilet material is collected at the children's school and composted on-site in wire bins, as shown above. The system is odor-free, yielding a compost excellent for growing plants.

Above and below: Compost bins for a group of households in Kampala, Uganda, are constructed of wood pallets on bare soil. The toilet material is collected from handicapped persons in the community. Below we see the bin feeding sequence: (1) establish the biological sponge; (2) add the toilet material and any other organic material such as food scraps and animal mortalities; (3) cover thoroughly with clean cover material. Repeat as needed until the bin is heaping full. The contents will shrink considerably as the process continues. Photos by Samuel Autran Dourado. **Opposite page:** Two young men fabricate compost bins from wire fencing in Tanzania, Africa. Photo by Joseph Lawrence Lacha.

(1) Keep the bin aboveground, stable, and vermin proof. Line it with wire mesh if you live in a rat-infested area. Or get cats.

(2) Start the bin on a soil base with a bowl-like bottom depression.

(3) Start with a thick "biological sponge" underneath the pile.

(4) Add your fresh material into a depression in the center of the bin contents. Put smelly organic material *in* the bin, not *on* the bin.

(5) Always cover the contents of the bin until there is no odor and there are no flies.

(6) Build a layer of cover material around the insides of the bin. This is facilitated by feeding new material into the center of the bin.

(7) Once the bin is full, let it sit, undisturbed, for about a year before using the compost. Always keep a clean layer of cover material on top of the compost pile. When you dig out a mature bin of compost, set aside the cover material that's on top and around the sides of the bin. Use this material for the biological sponge in the next bin.

(8) Don't try to rush compost. Immature compost kills plants. It doesn't cost anything to let the compost age and cure.

Discarded bagasse in Haiti covers acres of ground near a rum distillery.

Chapter Four

COVER MATERIAL

For a water toilet to function, you need water. For a compost toilet to function, you need a carbon-based, plant-derived cover material. If you don't have water, you won't have a water toilet. If you don't have carbon-based cover material, you won't have a compost toilet. A simple way to determine if something is carbon based is to ask yourself, if it's dry, will it burn? You can't use ashes, you can't use sand, you can't use lime, and you can't use dirt. None of these will burn because they lack the carbon. Most plant materials, on the other hand, if dry, will burn.

Cover materials successfully used around the world include sugarcane bagasse, which is ground and shredded sugarcane stalks used in the sugar and rum industries, found in most tropical climates. It contains residual sugar as well as cellulose, and microbes love it. Of course, sawdust can be found worldwide. The best is what comes from cutting trees into boards, beams, or posts, or from grinding trees to express oils. Sawdust is not wood chips, and it is not wood shavings. Chips come from a chipper, and they're

Continued on page 72

A variety of cover materials can be used around the world.

WOOD CHIPS, WOOD SHAVINGS, SAWDUST

Wood chips (above) are very difficult to compost and should never be used as a cover material. **Wood shavings** (below) are also difficult to compost but can be used in larger commercial compost piles. The shavings can also be left outside where they can get wet from rain or other liquid, becoming biologically activated, partially decomposed, and much more effective as cover material in compost toilets and bins. **Sawdust** works best when it comes from trees rather than from dry lumber, although dry sawdust can also be rehydrated and biologically activated.

Collecting biological sponge and cover material for the Sunflower House, Chiang Rai, Thailand. (Friends of Thai Daughters) Photo by Samuel Autran Dourado.

Volunteers collecting biological sponge and cover material in Kamwokya, Kampala, Uganda, Africa. Photo by Samuel Autran Dourado.

This compost bin, designed by the author, has a central section for storing straw bales for cover material. A roof over the straw keeps it dry so it won't freeze in the winter. The roof also collects rain water in a recycled wine barrel, which is used to wash toilet receptacles. The compost bins are on either side of the central bin.

If you can't find any cover material, make your own! Chipper/shredder machines can grind up unusable vegetation, such as the thorny blackberry brambles shown above, a fast-growing nuisance plant in California, USA. Even seemingly useless plants produce a suitable cover material for compost toilets. Some moisture content in the cover material will aid in odor control. Photos by Linn Harding.

too big for bacteria to eat. Shavings come from planing machines, and they also produce relatively large pieces of wood, which bacteria have a hard time consuming. In large municipal compost piles, wood shavings may work fine, given enough time. In smaller compost piles they will slow the pile down, especially if the wood was kiln dried. Allow the shavings to sit outside, get rained on, absorb moisture, become biologically active, and start to decompose. *Then* use them as cover material.

Rice husks or hulls, a by-product of the rice industry, are often used as cover material in toilets. They also tend to slow down the compost in smaller piles, but they do work. The by-products of cassava distilleries have been successfully used as cover materials when composting sewage sludge in China. Other promising cover materials in compost piles include olive mill by-products and sweet sorghum bagasse.

This compost bin wall was removed while the bin was being repaired, revealing how the cover material is enveloping the compost inside the bin. This is achieved by center feeding the bins, and it allows for compost to be collected in bins that are air impermeable, such as plastic, brick, block, or metal bins, assuming the bottom of the bin is open, and the compost is located on a soil base. When the finished compost is removed, the cover material envelope is used as the biological sponge for the next batch of compost.

You can produce your own cover material from almost any plant source by using a "chipper/shredder" if the machine can grind the material fine enough. The cover material doesn't need to be dry unless you're composting liquids. In fact, a small amount of moisture content is ideal.

Two Categories of Cover Material

There are two categories of cover material: (1) the cover material for *covering your toilet contents*, and (2) the cover material for *covering your compost piles*. They are not necessarily the same, although they can be. Inside the toilet you will need a finer material that has some degree of residual moisture. Sawdust from trees, with residual sap, is perfect. The residual moisture is what makes it an effective *biofilter*, or odor blocker. Bacteria live in the wet biofilms coating the wood particles. If you're using dry sawdust as a cover material and you notice odor from the toilet, mist the cover material with water when you're adding it to the toilet.

The straw/grass cover material is pulled aside to expose the compost underneath in a household compost bin in Tipitapa, Managua, Nicaragua. The contents of the compost bins, as well as the contents of the toilets, must always be kept covered with a suitable cover material to prevent odors and flies. This is a project of the NGO Sweet Progress.

If you use an appropriate cover material in adequate quantities, all odors will be blocked, and no flies will be attracted to the toilet or compost pile. This can't be emphasized enough. The cover material is the biofilter. It's the cover material that eliminates the need for venting. When you're using appropriate cover material, a standard toilet seat lid (other than the cover material) is all that is required to cover the toilet contents. The toilet receptacle itself never needs its own separate lid until it's removed from the toilet cabinet.

On the other hand, the cover materials used *on the compost pile* don't have to be in fine particles and can be either dry or moist. Straw works great. Hay is good, as are grasses, weeds, leaves, or anything from a plant source that is clean and doesn't smell bad. You don't want to use barnyard manures as cover material because they have unpleasant odors. Your toilet and your compost bin should be completely odor-free at all times. With proper management, they will be.

Cover materials help keep your pile aerobic by creating tiny interstitial air spaces in the compost. That's all the oxygen your compost will need. Large bulky materials are not needed in compost piles to create air spaces. We're talking about *microscopic* organisms. If the compost is aboveground and not under water, it will have air spaces in it. *Turning, digging, or chopping the compost is not necessary.* Stick a compost thermometer in your compost and keep an eye on it. If it's heating above ambient temperatures, your compost is active.

During cold weather, cover material stored outside can freeze solid, so it should be covered or insulated in some manner. Containers stored in a basement, heated garage, or enclosed porch will provide cover material throughout cold winter months.

Chapter Five

MANAGEMENT

The final necessary element in a compost toilet system is human management. You, or someone else, or maybe a team, must take responsibility for the compost toilet system. Receptacles must be cleaned, emptied, and available for use at all times. Cover material must be kept in constant supply, and the compost pile must be managed responsibly. It helps if the manager of the system is also someone who wants and values the compost. Making compost is creating something of value. There's not much point in making it if you're not going to use it.

Composting is both a science and an art. The point is to recycle organic material in an odor-free, nuisance-free, environmentally safe, and hygienically effective manner. Pay attention to what you're doing, and adjust your methods over time as needed. A compost toilet system will fail if managed poorly.

Public education is very helpful when introducing a compost sanitation system to a school, village, or other group setting. The fact that responsibly managed compost toilet systems are odor-free is hard to believe unless one sees for oneself. You can collect tons of compost, made from the toilet material of thousands of people, in dozens of open outdoor bins, and you can walk among the bins in hot tropical conditions without a whiff of anything unpleasant.

Children in a school in Nicaragua were taught to use the compost toilets brought to their school to replace their pit latrines. Then, they went home and taught their parents, who in turn abandoned either their own pit latrines, or their practice of open defecation, and adopted compost toilets in their homes. Without training and education, however, people would not know about or understand composting or compost toilets. One of the most common comments made by people when they see a compost toilet system for the first time is, "I never knew such a thing existed."

Managing a toilet is pretty straightforward. First, when a receptacle is filled, there must be empty ones nearby to allow for a quick exchange of toilet containers. Otherwise, the toilet will be unusable at times, much to the disappointment of people needing it. It's a simple matter to keep empty receptacles nearby, with lids, and with a few inches of cover material already in the bottom.

Second, there must be a steady supply of cover material within arm's reach of the toilet. The cover material should be finely ground and have some residual moisture content. The coarser and drier it is, the less likely that it will block all odor. Also, coarse, dry, woody materials do not compost very well.

Provide a device of some sort for scooping the cover material out of whatever container it is in. In public venues, such as at music festivals, for example, where inexperienced users may be using the compost toilet at night and potentially in the darkness, tie a cord or string to the scoop so it can't accidentally be dropped into the toilet. There's nothing worse than grabbing the scoop and finding that it has someone's poop on the handle because someone accidentally dropped it into the toilet.

Third, locate the toilet where it is most convenient. Just because your pit latrines are 50 meters away from your living space doesn't mean your

Community education teaches participants how and why to use a compost toilet in Nicaragua (opposite page), Haiti (above), and Mongolia (below). People who have used pit latrines all of their lives become excited to know that there is a private, secure, convenient, odorless toilet alternative within their economic means.

compost toilet has to be placed somewhere where it can't be quickly and conveniently accessed. In fact, compost toilets don't even need dedicated buildings. The toilet can be designed to be small and quite portable, as are many shown in this book. It doesn't take much space. Put it inside your house or school. Create a private area inside that is comfortable and pleasant. That way, when it's raining at night and you or your children need a toilet, there will be one readily available. One of the hardest concepts to grasp, for people in open defecation or pit latrine cultures, is that a toilet

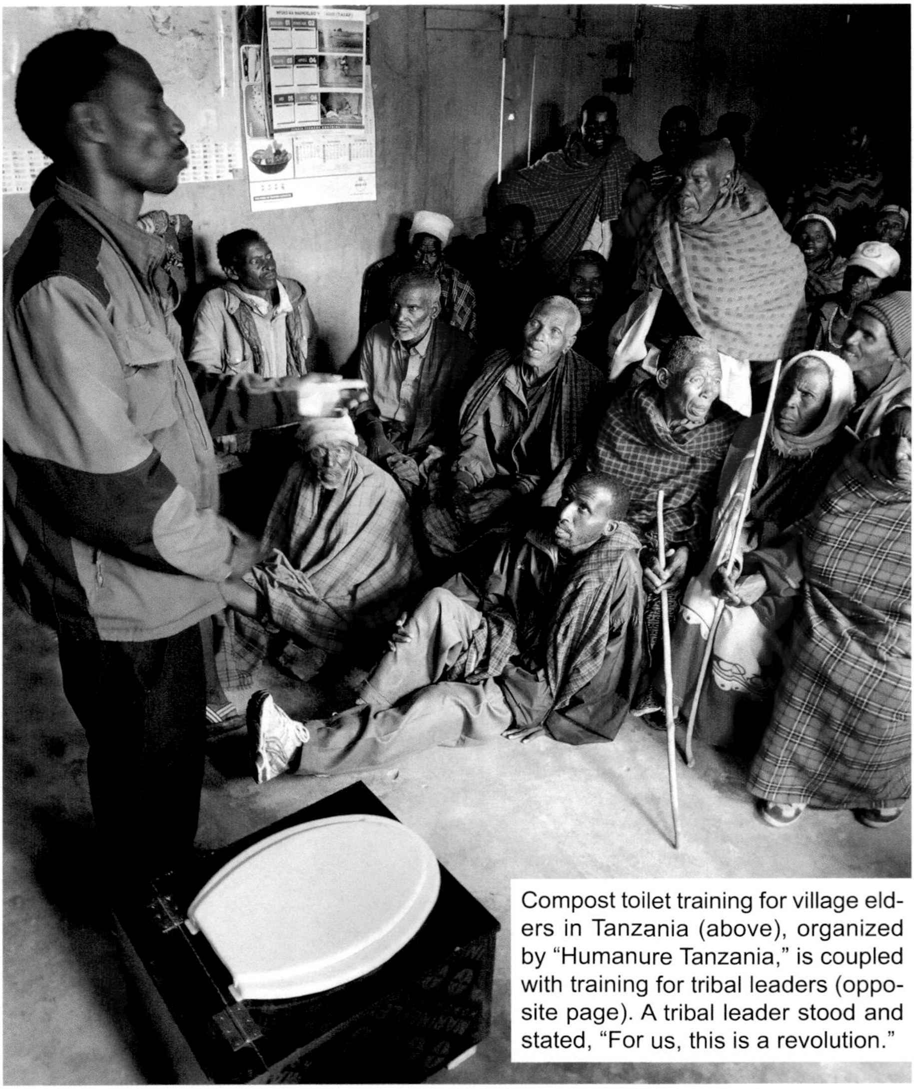

Compost toilet training for village elders in Tanzania (above), organized by "Humanure Tanzania," is coupled with training for tribal leaders (opposite page). A tribal leader stood and stated, "For us, this is a revolution."

doesn't have to smell or be inconvenient, and that it can be indoors and nearby. With the correct cover material, supplied in adequate quantities, and with sufficient toilet receptacles and *proper management*, a compost toilet can be indoors, pleasant, and convenient. A compost toilet does not need a vent, a drain, electricity, pipes, running water, urine separation, chemicals, or anything complicated or expensive. So why not put one in your office, bedroom, guest quarters, or school? This is especially important when considering elderly persons who may have problems walking or standing. A nearby toilet is essential.

Fourth, the toilet area must be kept clean. This is not difficult. The toilet *receptacles* are only cleaned at the compost bin or yard where the dirty water can be deposited into the compost bin, so you don't need to clean the inside of the toilet while it's in the toilet room. The messiest characteristic of a compost toilet is the cover material, which can become spilled or scattered about the toilet area, especially when children use the toilet. The solution is simple: keep a small broom or brush near the toilet.

Continued on page 84

Author's Field Notes from Tanzania, Africa

The tribal "meeting" turned out to be a gathering of Iraqw tribal leaders who walked "a kilometer or two" to this hillside tree to hear me explain to them what a compost toilet is and how it works. I spent an hour explaining this to them while Joseph [Lawrence Lacha] translated for me. Again, he did a great job and had everyone laughing, over and over. After explaining the nuts and bolts of using the toilet and making a compost bin, I asked Joseph to explain to them that half of all life on Earth is invisible because it's too small to be seen. And that compost toilets and composting are relatively new to humans because microbiology is new, and our understanding of invisible life is new and still developing. I told them that each of us has enough invisible life in and on our bodies that if we could scrape it all up into a pile, that pile would weigh as much as our brain. I told them that a lot of bacteria come out of our bodies in our excrement and that's how they get into the compost pile. With compost, we are able to tap into that invisible life and have it work for us by converting our organic material back into soil. I told them that a compost pile is a living thing, like a goat. It needs to be fed and cared for, and if we do so, it will help feed us by turning organic material into soil. I explained that even dead animals will be converted by a compost pile into soil that can be used to grow food. I added that this toilet is part of a knowledge-based system, a toilet for those who think, a "thinking person's toilet," a recycling system, not a waste disposal system. In the end, they were quite receptive to the idea of using these toilets, especially when they understood that the toilets can be indoors. The benefit to the infirm, the elderly, and the crippled was immediately obvious to them, even exciting.

I also explained to them that I have been using this toilet system for forty years in the USA, that I am not paid to be there in Tanzania, that I'm a volunteer, I cover my own expenses, and that I don't work for a government agency, academic institution, or NGO. Also, the toilet system is in use around the world. It is a waste-free toilet system that recycles organic material and protects groundwater supplies from the leachate that can come from pit latrines, and when you're carrying all your water from the river, protection of water supplies is extremely important.

So the obvious question was, how do they get this toilet system, how much does it cost, and who is going to pay for it? I explained to them that we were providing the first toilets free and that they should go look at them in use at the demonstration houses. I explained to them that they can also build their own toilets. It occurred to me standing there in front of this tribal group how important and essential it was to them to have this next revolution in sanitation and that we need funding to help them. I thought, "(insert billionaire's name here), you asshole, here is where you can do great help for a small amount of money." We need to design a compost toilet, maybe of plastic, that can be stacked and shipped in quantity, sold inexpensively, and mass produced. We also need to develop cover material technology and supply. . . .

Proper management of a compost toilet system requires that the system remain "closed." Cleaning water is used sparingly so as not produce too much soiled water. All of the soiled water, including soapy water, is deposited into the compost bins. Final rinse water can be dumped into a soak pit (hole filled with stones).

Photos by Samuel Autran Dourado taken at the Ayder Referral Hospital & medical school, Mekelle, Tigray, Ethiopia.

Above: A well-managed compost toilet system will have a person or persons in charge who checks the toilets on a regular basis, keeps the cover material supplied, makes sure full receptacles are swapped out for empties, adds cover material to the toilet when previous users had failed to use enough, makes sure the toilet area is clean, and makes sure toilet paper or cleaning water is available. **Opposite:** Community education is key for the adoption of compost sanitation systems in communities where they have never been used, or even heard of. Photos by Samuel Autran Dourado (opposite page in Torit, South Sudan, a NIRAS project; and above at Police Academy Children's School in Moroto county, Karamoja, Uganda, a WeltHungerHilfe project).

The toilet receptacles, when filled, can be set aside, with lids, and taken to a compost bin when it is convenient. Using only one receptacle and emptying it when it's filled is a mistake, because you always want empty ones available to immediately "swap out" the full ones. Provide indoor space to allow them to accumulate if needed. During a snowstorm, hurricane, or other bad weather you don't want to be emptying compost containers if you don't have to. It's up to you to make the system convenient and successful. This may take a little bit of thought and planning at times. Bad weather coming? Take care of your compost before it arrives, then sit back, relax, and let it pass while you and your family enjoy the benefit of an odorless indoor toilet. Power goes out? No electricity? No running water? No matter. The toilet will still function. When the weather breaks, take care of your compost again. Just make sure you have enough receptacles to last for whatever period of time you need.

Keep the toilet room supplied. If you use toilet paper, keep it stocked and readily available, so the toilet users don't run out. If you use water to wash your bum, collect the dirty wash water inside the toilet and use extra cover material to absorb the extra liquid. Decorate the toilet room. Make

it a pleasant place for a short rest. Air fresheners, lights, reading material, hand sanitizer, and additional accoutrements may be a nice touch.

Insect problems? Insects are rarely a problem, although fruit flies and gnats can become a nuisance if you put food materials inside your toilet, or if your cover material somehow has eggs or insects already in it. If you leave a used toilet receptacle unattended inside a toilet cabinet for a long period of time, especially if there is a food item in it such as a fruit core or peel, you can expect trouble. If you set your toilet receptacles aside for later composting and add food material but don't keep a snug lid on it, or if a lid becomes accidentally dislodged, you're asking for insects to show up.

Don't put food scraps into a compost toilet receptacle unless it has been removed from the toilet cabinet and has a snug lid on it. Then you can add whatever you want. If you have a gnat or fruit fly problem, cover the toilet contents thoroughly and remove the problem toilet receptacle from the toilet area. Put a snug lid on it. Replace it with a fresh receptacle. Make natural insect spray from soapy water and an essential oil, such as peppermint or clove oil, and spray it inside your toilet as you open it. If there are gnats in there, they will soon disappear. Plus, the spray smells nice. Use a bottle that sprays a fine mist, squirt some dish soap into it, add several drops of essential oil, then fill it with water. Keep the bottle in your toilet room. If you're having a persistent issue with flying insects, hang some sticky fly paper somewhere in the room to catch the strays. It works. One can only guess what kind of insects you may have to deal with in your location, wherever you are in the world. Good cover material, proper management, and maybe some creative thinking will ensure success.

Larger scale compost toilet systems, such as a managed village system, a music festival, a school, an orphanage, or a community, may require a team effort. One or more persons may choose to keep the toilets clean and supplied. Others may focus on keeping the cover material in sufficient supply. Still others may want to manage the compost itself. There are several case studies reviewed in the second half of this book that will shed some light on how these larger systems work.

Image on opposite page modified from an original diagram by Gabriela Veras.

REOTEMP
Compost
Steady
Hot
°F
Thermometer

Chapter Six

COLD WEATHER COMPOSTING

The author lives in an area of the US where the temperature has dropped as low as –30 °F (–34 °C), and snow can fall from late October into May. How do you make compost in cold weather?

There are a number of things to take into consideration when managing compost during cold winter months. One is to have a heated place to store both toilet cover material and full toilet receptacles. The filled receptacles can be collected, set aside with lids, then emptied outside in a compost bin *when weather permits*. During cold winter months in extreme climates, you may find it advantageous to accumulate toilet containers until there is a break in the weather. You will probably also have to carry out unfrozen, or preferably warm, water to the compost bins with which to clean the receptacles. Otherwise, line the receptacles with compostable plastic bags to eliminate having to wash them after emptying.

It's important to not let full toilet receptacles freeze because the contents will expand and potentially split open the container and cause leakage. Also, if a full receptacle becomes frozen, it's almost impossible to empty it. You will have to thaw it out before you can empty it, and then hope it's not cracked and going to leak!

Keeping a thick cover material on top of the compost will help insulate it. Surrounding the compost with an insulating layer of cover material helps, too. Bin sidewalls could be made of straw or hay bales, which would act as insulation in colder climates. After the compost has matured inside the straw walls, the bales can be used as the cover material in the next bin.

Regular feeding of a compost pile will help keep it active and warm

Opposite page: The bottom photo of the author's household compost bin in northwestern Pennsylvania USA was taken December 20, 2020, with only standard household organic material, including toilet material, in the bin. The compost is 120°F (49°C). The top photo is in January (year uncertain). The center bin keeps rain and most of the snow off the straw bales. Wet bales will freeze solid and be unusable in the winter. A flat "shed roof" sloped to the front or back would also keep snow out of the bins. The gable roof shown has "snow guards" attached to the roof to prevent the snow from sliding into the bins off the roof. Too much snow in the bins makes it difficult to access the compost.

The author's household compost pile in the dead of winter can remain hot if it is continuously fed (at least weekly). Hot food-grade liquids will stimulate compost during cold weather, too. The outdoor temperature when this photo was taken was 4 °F (–15 °C). It's more common, however, for compost piles to have reduced temperatures during cold winter months. They may even freeze. Although the author's compost piles initially froze annually for years, they thawed out and heated up again every spring.

during cold weather. Food-grade liquids added to the pile will also help keep it active, especially if the liquid is at room temperature, or warmer, even quite hot. The author has found that hot, "spent" wine, a by-product of the brandy distillation process, is favored by compost microbes in the winter, as is room temperature discarded beer overflow from breweries.

An important concern is having unfrozen cover material readily available during freezing weather. The author puts two large, wheeled, plastic garbage cans filled with sawdust into an enclosed porch during the winter months and keeps his straw bales dry in a roofed compost bin. The straw makes a convenient cover material in the bin. The author's compost hasn't frozen in over twenty years, despite the extreme weather.

Theoretically, the microbes in the author's compost, having multiplied for over forty years in the same location, may have adapted to the climate to be more resistant to the cold weather. Residual bacteria can reinoculate new compost piles from the soil base underneath them.

Standing Rock

A 2017 pioneering cold-weather compost toilet project was conducted in rural North Dakota, USA, by GiveLove (Los Angeles, California, USA; GiveLove.org). The project at the Standing Rock protest was coordinated by lead project manager Alisa Puga Keesey, who presented her data at the 2018 World Dry Toilet Conference in Finland (*Container-based Sanitation at the Standing Rock Protest: an Experiment in Communal Toilet Management and Extreme Composting*).

GiveLove is a US-based skills training nongovernmental organization dedicated to the teaching and promotion of ecological sanitation. Since 2010, the organization has worked to introduce compost-based sanitation systems for high-need and water-scarce areas throughout the world.

Thousands of people from all over the world traveled to the Standing Rock Indian Reservation to join the historical protest against the construction of the Dakota Access Pipeline. As the camps swelled to over fifteen thousand people at their peak, the Standing Rock Sioux Tribe struggled to provide clean toilets for everyone. A company was hired to provide 120

portable toilets at a cost of over $30,000 US per month.

Native activists called "water protectors" were concerned about the environmental pollution caused by the chemical toilets, which were being discharged on the nearby prairie. The portable toilets were also inoperable when frozen, and it was urgent to find an alternative. Hoping to avert a public health and environmental crisis, the tribe approved GiveLove's proposal to introduce a compost toilet system in the camps. The objectives included: 1) clean "dry" toilets that would function in winter conditions; 2) an environmentally sustainable system; and 3) on-site treatment of the toilet material on the reservation.

GiveLove began building compost toilets at the camp in November 2016. A location for a permanent composting site on reservation land, approximately 5 kilometers from the camps, was identified and evaluated by an environmental engineer. Work proceeded at a frenetic pace as the team raced to launch the project before the arrival of winter snowstorms. The main objective was to provide clean compost toilets and comfortable toilet facilities for the thousands of activists residing in the camps, including families with small children, and elders.

The project was budgeted at $156,000 US for twelve months, which included the construction of the toilet blocks and the compost site. Skilled labor was recruited for the construction phase, which included the retrofitting of six 6 x 12-meter US Army tents. Five toilet blocks contained thirteen toilets per tent, including stalls designated for LGBT and disabled persons. The tents included flooring, lighting, storage areas for toilet supplies, and wood stoves for heating. Two shipping containers were used to hold the frozen toilet material until it could be composted. Volunteers were organized and trained for the daily management of the compost toilets, cleaning and maintenance of the toilet facilities, logistical management of the operation and supplies, and collection of the toilet material. Toilet use demonstrations were conducted daily. Volunteers worked eight-hour service shifts twenty-four hours a day, seven days a week. This in-

Opposite page: The Standing Rock organizers included Alisa Keesey, far left, and Patricia Arquette (holding the toilet seat). Steffan Thimmes, the trainer, is standing behind Patricia. They are aided by "water protector" volunteers. Note the toilet tents behind them. Photo provided by GiveLove.org.

Compostable plastic bags filled with toilet material are being layered with straw and silage (above). Although fifteen thousand people were serviced at the peak of the gathering, and weather was bitter cold at times, innovation and perseverance prevented thousands of gallons of sewage from being created, instead producing tons of compost.
Photos provided by GiveLove.org.

A Compost Toilet at Standing Rock

Compostable bags made by the company BioBag™ were used to line the 20 liter (5-gallon) containers.

The containers were also lined with a regular plastic trash bag to prevent the compostable bag from sticking to the sides.

With two bags as liners, no water was needed to wash the containers.

Fine wood savings for the toilet were purchased at a local feed store for US$5.00 per bag.

cluded taking the full bags of toilet material to the storage site.

The toilet tents had to be warm enough to prevent the toilet material from freezing, so there were wood stoves burning 24/7. Because there was no water for cleaning toilet receptacles, 13-gallon compostable plastic bags were used to line the toilets. However, the bags stuck to the plastic 20-liter toilet receptacles due to the cool, humid conditions. Conventional plastic trash bags were therefore taped inside the toilet receptacles, allowing the BioBags to easily slide out of the containers when full, and for easy clean-up. Three full BioBags, weighing approximately 4 to 6 kilograms each, were stored inside large plastic bags outside of the tents where they froze solid each night. The bags were transported to the shipping containers every two to three days. No urine diversion was needed. All urine was collected in the toilet receptacles along with all fecal material.

The toilet blocks quickly became the central meeting place in the camp when the cold weather prevented people from socializing outside. The toilet blocks were warm inside, and they didn't smell at all. People met there with friends, started playing music and movies, and decorated the walls with art, Christmas lights, and posters. Toilet duty became the "best job in the camp," and toilet tents became a fun place to hang out in, while hundreds of people voiced gratitude for the nice, clean toilets.

The compost bins, constructed with straw bales reinforced with rebar for stability, had biological sponges made of spoiled hay. The compost team removed the semi-frozen bags of toilet material while wearing protective gear. Spoiled silage and hay were sourced from local ranches as a compost feedstock, in addition to the straw bales. The BioBags were piled inside the compost bins. Approximately two thousand bags of toilet material were added to the compost bins over six days. The organic material was left in the bins for eighteen months. Temperatures were monitored regularly.

The project at Standing Rock was most likely the largest emergency compost toilet system ever implemented in the United States, and the first of its kind in an extreme climate and setting. The success of the sanitation system confirmed the adaptability of compost toilets in emergency situations under less-than-ideal circumstances.

The compost above was made at a school in Haiti from toilet material and sugar cane bagasse. It has a rich and loamy appearance and a pleasant aroma.

Chapter Seven

WHAT ELSE CAN COMPOST DO?

Compost microorganisms not only convert organic material into compost and eliminate disease organisms in the process, but they also degrade toxic chemicals into simpler, benign, organic molecules. These chemicals include gasoline, diesel fuel, jet fuel, oil, grease, wood preservatives, polychlorinated biphenyls (PCBs), coal gasification wastes, refinery wastes, insecticides, herbicides, TNT, and other explosives.

Compost can also be used to control odors. Biological filtration systems, or "biofilters," are used at large-scale composting facilities where exhaust gases are filtered for odor control. The biofilters are composed of layers of

Festering garbage piles are an unfortunate sight in many parts of the developing world. They breed rats and flies and produce a strong, unpleasant stench. The contents of these piles consist primarily of organic material that can be cocomposted with toilet material, thereby eliminating the garbage and the sewage at the same time and replacing them with a pleasant-smelling humus suitable for gardening or agriculture.

Compost workers add toilet material to a bin at the Kwale's girl's high school, Kwale, Kenya (an Aqua-Aero WaterSystems project). Photo by Samuel Autran Dourado.

organic material such as wood chips, peat, soil, and compost, through which exhaust air is drawn to remove contaminants from the exhaust. The microorganisms in the organic material "eat" the contaminants and convert them into carbon dioxide and water.

Compost microbes directly compete with, inhibit, or kill organisms that cause diseases in plants, too. The microbes can also produce antibiotics that suppress plant diseases. Because of this, diseased plant material should be composted rather than returned straight to the land, where reinoculation of the disease could occur. Plant pathogens are also eaten by microarthropods, such as mites and springtails, which are found in compost.

Compost added to soil can activate disease resistance genes in plants, providing them with a better defense against plant pathogens. Studies indicated that compost inhibited the growth of pathogenic microorganisms in greenhouses because it added beneficial microorganisms to the greenhouse soils. The studies suggested that sterile soils could provide optimum breeding conditions for plant disease microorganisms, while a rich diversity of microorganisms in soil, such as that found in compost, would render the soil unfit for the proliferation of disease organisms.

Compost tea has also been demonstrated to have disease-reducing properties in plants. It is made by soaking mature, but not overly mature, compost in water for three to twelve days. The tea is then filtered and sprayed on plants undiluted, coating the leaves with live bacteria colonies.

When scientists inoculated wood chips with three different fungal plant pathogens, then composted them, they found that a temperature of 104°F (40°C), when exceeded for more than five days, was sufficient to kill all three of the organisms. Composting also destroys weed seeds. Researchers observed that after three days in compost at 131°F (55°C), all the seeds of the eight weed species studied were dead.

Composting is considered a simple, economic, environmentally sound, and effective method of managing animal mortalities. Dead animals of all species and sizes can be recycled, including full-grown pigs, cattle, horses, fish, sheep, and calves. The composting process not only converts the carcasses to compost that can be applied directly to farmers' fields, but it also destroys the pathogens and parasites that may have killed the animals in the first place. A temperature of 131°F (55°C) maintained for at least three consecutive days maximizes pathogen destruction.

Carcasses should be buried in a compost pile and not in the ground where they may pollute groundwater, as is typical when composting is not used. Carcass composting can be accomplished without odors, flies, or scavenging birds or animals. Make sure your compost bin has animal-proof sidewalls, then simply lay a piece of stiff wire fencing on top of the compost to prevent animals from digging into it.

Extra care and diligence must be used when composting animal carcasses, because the stench is awful unless the carcasses are *thoroughly* buried. Toilet material is great for directly covering carcasses — ironically, it blocks

This Nicaraguan lady's father tragically lost both of his legs. Benefactors installed a flush toilet in their house and a septic system for his benefit, but there was no reliable water source available to flush the toilet. The only usable toilet was a pit latrine quite some distance away. Now the family has a convenient compost toilet and a nearby compost bin. No water is needed other than a small amount to wash the toilet receptacles.

the odors! Add green material or food scraps too when covering an animal carcass. If you still smell something, add more of the same until the odor is completely blocked!

Woods End Laboratories in Maine did some research on composting ground-up telephone books and newsprint that had been used as bedding for dairy cattle. The ink in the paper contained common cancer-causing chemicals, but after composting it with dairy cow manure, the dangerous chemicals were reduced by 98 percent. So it appears that if you're using shredded newspaper for animal bedding, you should compost it, if for no other reason than to eliminate toxic elements from the newsprint.

Sewage, waste, pollution, garbage, and all of the ills associated with them can be converted into wholesome, edible, human sustenance by composting the organic material, adding the compost to the soil, planting seeds, tending the crops, harvesting the food, and learning how to preserve and prepare the bounty.

COMPOSTING ANIMAL MORTALITIES

(1) Rake aside the cover material and dig a hole in the compost pile.

(2) Drop the carcass(es) into the hole.

(3) Pour toilet material over the carcass, then green material, if available.

(4) Pull the compost and the cover material over the new deposits.

(5) Add fresh cover material on top. Keep a piece of wire fence on top of the pile if you have problems with animals digging into it.

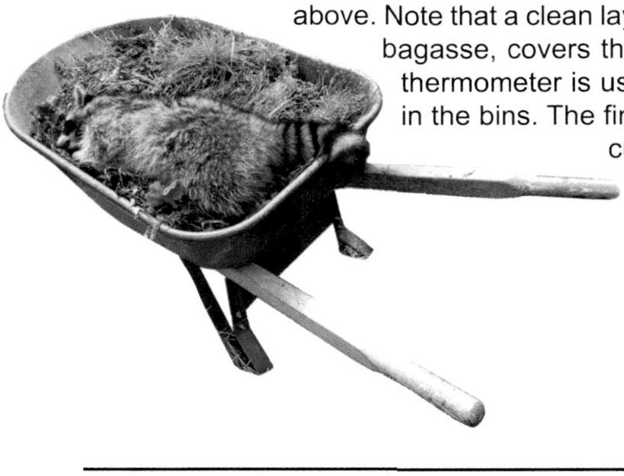

Toilet material from a school in Haiti is collected in bins by the compost manager, above. Note that a clean layer of cover material, in this case sugarcane bagasse, covers the bin contents at all times, and a compost thermometer is used to monitor the level of microbial activity in the bins. The finished compost is a pleasant-smelling agricultural soil conditioner that is safe to handle and can be stored indefinitely.

Chapter Eight

MAKING AND USING COMPOST

One thing rarely thought of when toilets are mentioned is *gardens*. Sanitation professionals have spent their careers trying to dispose of, or "treat" "human waste." Ironically, it is the sanitation authorities who are *creating* the waste by insisting that toilet material must be disposed of rather than recycled. They often *demand* that the "waste" be dumped into water via a flush toilet, thereby guaranteeing the creation of sewage and the pollution of our water supplies. This is the status quo for flush toilet cultures, which are then faced with the task of trying to clean up the rivers, lakes, streams, groundwater, and estuaries when sewage manages to seep into them.

Little concern is given to the loss of soil fertility that is the by-product of waste production. Imagine what it would be like if a toilet was not a waste disposal device, but instead something that made gardens. Imagine if the sewage, waste, and pollution typically associated with toilets was somehow magically replaced with food for the kitchen table. Imagine if somehow those rank, smelly turds your body excretes on a daily basis could be converted into tomatoes, peppers, peaches, potatoes, squashes, and scores of other edible plants and their fruits. If a shudder of revulsion is coming over you at the thought of such a thing, you are not alone. However, toilets *can* make gardens and can do so easily, at low cost, very simply, without electricity, water, chemicals, or even odors. Furthermore, such toilets, *compost toilets*, are within the reach of people of the lowest economic means.

This chapter contains photos of the author's organic garden, which, at the time of this writing, had benefited from forty years of humanure compost. All of the organic material collected in the household compost toilets was recycled by composting, and almost all of the compost was added to the garden soil. Some was used for trees, shrubs, and houseplants. That compost included many other "feedstocks" (things fed to the compost pile), such as kitchen scraps, garden residues including weeds and prunings, grass clippings, leaves, animal mortalities, liquid food byproducts, and so on. The photos were taken at various times throughout the first forty years, during which no animal manures were used other than the manure from the author's few chickens, which was primarily used as a mulch.

Human excrement is readily compostable; it comes from our bodies and exists inside us at all times. It is teeming with beneficial microorganisms and is not something to be feared. According to an American MD, whose family uses a compost toilet, "There is no offensive odor. We've never had a complaint from the neighbors." His brief description sums it up: "This simple compost toilet system is inexpensive both in construction and to operate and, when properly maintained, aesthetic and hygienic. It is a perfect complement to organic gardening. It outperforms complicated systems costing hundreds of times as much."

Well-informed health professionals and environmental authorities are aware that "human waste" presents an environmental dilemma that is not going away. The problem, on the contrary, is getting worse. Too much land and water are being polluted by the sewage and septic discharges of too many people, and constructive alternatives must be explored.

Compost Basics

Composting is a continuous natural process whereby organic material is recycled to make soil for plants. The plants produce food that is eaten by animals, including humans. The animals then produce manures, carcasses, discarded food, and other agricultural residues. When the manures, discarded food, and other organic by-products are composted, the cycle continues, without waste.

Organic material that will make compost could be anything that had been alive, or from a living thing, such as manures, plants, leaves, sawdust, peat moss, straw, grass clippings, food scraps, and urine. Anything that will rot will compost, including such things as cotton clothing, wool rugs, rags, paper, dead animals, junk mail, and cardboard. Composting converts organic material, including toilet material, into a stable substance that does not have odors, and does not attract insects or nuisance animals. Mature compost can be safely handled and stored indefinitely.

Compost increases the soil's capacity to absorb and hold water. Compost also adds slow-release nutrients essential for plant growth, creates air spaces in soil, helps balance the soil pH, darkens the soil (thereby helping it to absorb heat), and supports microbial populations that add life to the soil. Nutrients in compost, such as nitrogen, are slowly released throughout the growing season, making them less susceptible to loss by leaching than more soluble chemical fertilizers.

Organic matter from compost enables the soil to immobilize and degrade pesticides, nitrates, phosphorus, and other chemicals that can become pollutants. Compost also binds pollutants in soil systems, reducing their leachability and their absorption by plants. By composting our discarded organic material and returning it to the land, we can restore soil fertility in relatively short periods of time. Fertile soil yields better food, thereby promoting better health.

Pile the compost aboveground in a bin. A pile contained in a bin (versus an open pile or windrow) keeps the organic material from drying out or cooling down prematurely. A level of moisture of 50 to 60 percent is necessary for the compost microorganisms to work effectively. An enclosed

The author's garden has benefited from decades of compost made from, among other things, humanure. Most toilets produce sewage. Compost toilets create gardens.

pile helps to prevent leaching and waterlogging, and it holds biological heat. A neat pile looks good in your backyard or in your community, instead of looking like an open dump. Compost in bins, as opposed to open piles, also keeps out nuisance animals such as dogs and goats. A bin doesn't have to cost much money; it can be made from recycled wood, cement blocks, hay bales, repurposed pallets, or whatever else is available.

Large-scale municipal composting is often done in windrows, which are long open piles that are usually uncovered. These open piles must be frequently turned and stirred because the exposed surfaces emit odors, attract flies, and can't heat up like the interior does. You will find that commercial large-scale windrow operations will often not accept animal manures, certainly not toilet material, and often not even food scraps, due to the odor and fly issues these feedstocks will create when the uncovered piles are sitting and rotting in the sun or are being stirred up and releasing gases and a host of other things into the air. The good news is that smelly feedstocks such as humanure, dead animals, and food materials can be composted in contained, covered piles that do not need to be stirred at all, thereby eliminating the odor and fly issues completely. A contained "static" composting technique also eliminates the considerable cost and labor involved in turning compost piles.

Turning the Piles

What is one of the first things that comes to mind when one thinks about compost? Turning a compost pile? A large industry has emerged from this philosophy, one that manufactures expensive compost turning equipment. A lot of money, energy, and expense go into making sure compost is turned regularly. For some compost professionals, the suggestion that compost doesn't need to be turned at all is ridiculous. Yet the perceived need to turn compost is one of the myths of composting.

Why do people turn compost piles? First, turning is supposed to add oxygen to the compost pile, which is supposed to be good for the aerobic microorganisms. Second, turning the compost ensures that all parts of the pile are subjected to the high internal heat, thereby ensuring total pathogen death and yielding a hygienically safe, finished compost. Third, the more we turn the compost, the more it becomes chopped and mixed, and the better it looks when finished, rendering it more marketable. Fourth, frequent turning can supposedly speed up the composting process.

If you don't market your compost, and you don't care if it's finely granulated or somewhat coarse, and you have no reason to be in a hurry, let's eliminate the last two reasons for turning compost and look at the first two.

Oxygen is necessary for aerobic compost, and there are numerous ways to aerate a compost pile. One is to force air into or suck air through the pile using fans, which is common at large-scale composting operations where air is pulled from under the compost piles and pumped out through a biofilter. The suction causes air to seep into the organic mass through the top, keeping it aerated. Such mechanical aeration is never a need of the household or village composter and is limited to large scale commercial composting operations where the piles are so big, they can smother themselves if not subjected to forced aeration.

Some degree of aeration might be achieved by poking holes in the compost and otherwise impaling it. Many composters are taught to physically turn or dig their piles. However, none of these practices are necessary. Instead, the object is to build the pile in a bin aboveground allowing tiny interstitial air spaces to be trapped in the compost. This is done by using materials in the compost such as hay, straw, weeds, and the like. When a compost pile is properly constructed aboveground, no additional aeration will be needed.

Composting is easy if you let the microbes do the work, especially if you're composting smelly things such as toilet materials, dead animals, and food scraps, all of which will stink and attract flies, rats, and dogs if you don't manage your compost correctly. The trick is always to keep the compost vertically contained in a vermin-proof bin, approximately a cubic meter in size or larger, and keep it covered. If it's aboveground and not under water, the organic mass will be aerobic. No forced aeration will be needed, and no poking, prodding, digging, or turning is required.

Put a biological sponge underneath the pile, then create a layer of cover material surrounding the composting mass. Always keep a substantial layer

One of the most difficult aspects of compost sanitation training is getting people to understand that they will be producing plants and food, and not waste.

of cover material on top of the pile, too. When adding fresh material to your compost bin, pull aside the cover material, dig a hole in the underlying compost, deposit the new material in the hole, cover it with the existing compost, pull the cover material back over, then add more cover material as needed. This feeds fresh material directly into the center part of the pile, leaves no exposed surface areas, and eliminates odors, flies, and any need to turn or agitate the pile. It's that simple.

If you're composting material from toilets, you don't want to be turning the pile if you don't have to. It is a huge amount of work to turn compost

When we recycle organic material by composting, we can maintain soil fertility. The author's compost has been used in his food garden for over forty years.

piles, so why do it if it's not necessary? Making compost can be like baking bread — you put it in the oven, let it cook, and don't touch it until it's completely done, then you pull it out when it's ready. When making compost, you're putting organic material into a bin and enveloping it with cover material, like putting bread dough in a bread pan and slipping it into an oven. Then, when the bin is full, you wait. When the compost temperature drops down to the ambient outdoor temperature, and not before, *then* you can take the bread out of the oven. Generally speaking, once the pile is completely built, wait *about* a year before harvesting the compost (give or take

Using compost in a garden is not using "night soil" or human excrement to grow food. Humanure is instead fed to microorganisms, which convert it to compost.

a couple of months). There are various other things to take into consideration — the compost pile shouldn't be too high, for example, or the compost will smother under its own weight.

Researchers have measured oxygen levels in large-scale windrow composting operations and reported, "Oxygen concentration measurements taken within the windrows during the most active stage of the composting process, showed that within fifteen minutes after turning the windrow — supposedly aerating it — the oxygen content was already depleted." Other researchers compared the oxygen levels of large turned and unturned batch compost piles and have come to the conclusion that compost piles are largely self-aerated. "The effect of pile turning was to refresh oxygen content, on average for [only] 1.5 hours (above the 10% level), after which it dropped to less than 5% and in most cases to 2% during the active phase of composting. . . . Even with no turning, all piles eventually resolve their

There is no need for a toilet to create odor, draw flies, or produce waste and pollution. It can instead produce fertile soil, beauty, abundance, food, and healthy people.

oxygen tension as maturity approaches, indicating that self-aeration alone can adequately furnish the composting process. . . . In other words, turning the piles has a temporal but little sustained influence on oxygen levels." These trials compared compost that was not turned, bucket turned, turned once every two weeks, and turned twice a week.

Interestingly, the same trials indicated that bacterial pathogens were destroyed whether the piles were turned or unturned, stating that there was no evidence that bacterial populations were influenced by turning schemes. There were no surviving *E. coli* or *Salmonella* strains, indicating that there were "no statistically significant effects attributable to turning."

The more frequently compost piles are turned, the more agricultural nutrients they lose. When the finished compost was analyzed for organic matter and nitrogen loss, the unturned compost showed the least loss. The more frequently the compost was turned, the greater was the loss of both nitrogen and organic matter. Also, the more the compost was turned, the more it cost. The unturned compost cost $3.05 US per wet ton to produce, while the compost turned twice a week cost $41.23, a 1,352 percent increase. The researchers concluded that "composting methods that require frequent turning are a curious result of modern popularity and technological development of composting as particularly evidenced in popular trade journals. They do not appear to be scientifically supportable based on these studies. . . . By carefully managing composting to achieve proper mixes and limited turning, the ideal of a quality product at low economic burden can be achieved." Another study concluded that the turning frequency of yard refuse windrow compost did not improve aeration, had little impact on temperatures, and increased the bulk density of the compost, which actually reduces oxygen availability.

When large piles of compost are turned, they give off emissions of such things as *Aspergillus fumigatus* fungi, which can cause health problems in people. Aerosol concentrations from static (unturned) piles are relatively nonexistent when compared to mechanically turned compost. Measurements 30 meters downwind from static piles showed that aerosol concentrations of *A. fumigatus* were not significantly above background levels and were "33 to 1800 times less" than those from piles that were being moved.

Gardens add pleasure to life. Gardening provides exercise, fresh air, creativity, "grounding" with the Earth, peace, tranquility, and ecological harmony, all of which enhance human health and well-being. Need something to keep you sane in this troubled world? Plant a garden, harvest the food, learn how to preserve it, and let it nourish you.

A clean cover material such as hay, straw, leaves, grass, or something similar can be used to cover compost piles. It can be removed later and used to make the next batch of compost. A generous cover material and no agitation of the pile will reduce odor emissions to zero, thereby eliminating the negative health effects of inhaling compost bioaerosols. Turning compost piles in cold climates can also cause them to lose too much heat. What's the point of diminishing the heat of the pile, no matter what the climate? That's what turning the pile does. That's why there are large clouds of vapor billowing off large compost piles as they're being turned by machines. It's the heat, and lots of other things, escaping into the air.

Moisture

Compost must be kept moist. It's amazing how much moisture an active compost pile can absorb. A compost pile is not a pile of garbage or waste. Thanks to the miracle of composting, the pile of organic material becomes a living, breathing, biological mass, a sponge that absorbs quite a lot of liquid. Due to compost's need for liquid, it is more likely that one will have to *add* moisture to one's compost than have to deal with excess moisture leaching from it.

More importantly, microorganisms don't have legs like land animals do, and they need moisture for motility. Microbes live in biofilms coating the particles and surfaces in a compost pile. When the compost dries out, biological activity slows down and eventually grinds to a halt.

The water required for compost making may be 200 to 300 gallons (757–1135 liters) for each cubic yard (0.76 cubic meters) of finished compost. This moisture can be acquired from urine and rainfall. Liquid can also come from food scraps. If adequate rainfall is not available and the contents of the pile are not moist, such as in a desert situation, watering may be necessary.

Gardening adds a spiritual element to life unhampered by the faults associated with religions. Sometimes there is no place closer to heaven on Earth than in a garden.

Carbon and Nitrogen

A good "carbon/nitrogen balance" is required for a successful compost pile. Since most of the materials commonly added to a backyard compost pile are high in carbon (leaves, for example), a source of nitrogen must be used in the blend. This isn't as difficult as it may seem: add manure.

By using all the organic refuse a family produces, including toilet material, food scraps, weeds, and grass clippings, with some materials from the larger agricultural community such as straw or hay, and maybe some rotting sawdust or some collected leaves, one can get a good mix of carbon and nitrogen for successful composting.

A good C:N ratio for a compost pile is between 20:1 and 35:1. That's 20 parts of carbon to 1 part of nitrogen, up to 35 parts of carbon to 1 part of nitrogen. Or for simplicity you can figure on shooting for an optimum 30:1 ratio. You can think of carbon as something that originates from plants and that will burn if dry. Ashes don't burn; they're what's left after burning, so there's no carbon there. Rocks don't burn, so limestone is not a carbon source. Most agricultural or natural plant residues, if dried out, will burn. Those are your carbon sources.

For microorganisms, carbon is the basic building block of life and is a source of energy, but nitrogen is also necessary for such things as proteins, genetic material, and cell structure. Microorganisms that digest compost need about 30 parts of carbon for every part of nitrogen they consume. If there's too much nitrogen, the microorganisms can't use it all and the excess is lost in the form of smelly ammonia gas. Nitrogen loss due to excess nitrogen in a compost pile (a low C:N ratio) can be over 60 percent. At a C:N ratio of 30 or 35 to 1, only 0.5 percent of the nitrogen may be lost. If you have a high-nitrogen feedstock such as humanure, chicken manure, urine, and so on, just add more carbon. How much? Enough that you can't smell anything — it really is that simple. Use your nose; it's a great tool!

Toilet material is too wet with too much nitrogen and not enough carbon to compost. Hay, straw, weeds, or even paper products if ground to the proper consistency will provide the carbon. Food scraps are generally already C:N balanced, so they can be readily added to your compost pile.

Sawmill sawdust has a moisture content of 40 to 65 percent, which is good for compost. Lumberyard sawdust, on the other hand, is kiln-dried and is biologically inert due to dehydration. Therefore, it is not as desirable in compost unless rehydrated with water or urine from your compost toilet before being added to the compost pile.

What about sanitary napkins and disposable diapers? Sure, they'll compost, but they may leave strips of plastic in your finished compost. That's okay if you don't mind picking the strips out of your compost later. Otherwise, try reusable cloth diapers and washable cloth menstrual pads.

Add compost to the holes when you're planting young trees to aid in new root growth. This "Rare Ripe" peach tree was started from a peach seed in the author's garden.

Fecal material can be scraped off cloth diapers into the compost toilet with toilet paper or leaves. The diapers can then then be soaked in a "diaper bucket" in water, eventually wrung out, laundered, and reused. The soiled water from the diaper bucket can be dumped into the compost pile.

Toilet paper composts, too. So do the cardboard tubes in the center of the rolls. There are some things that don't compost very well: eggshells, bones, and hair to name a few. But these things won't hurt your compost pile. Throw them in.

A section of the author's garden just before the first fall frost.

Phases of Compost

There is a big difference between a small family or village composter and a large-scale commercial composter. Municipal composters handle large batches of organic materials all at once, while household composters continuously produce small amounts of organic material every day. Municipal composters, therefore, could be termed "batch" composters, while backyard composters would be "continuous" composters. When organic material is composted in a batch, four distinct stages of the composting process are apparent. Although the same phases occur during continuous composting, they are not as apparent as they are in a batch, and in fact they may be occurring concurrently rather than sequentially.

The four phases include: (1) the preliminary mesophilic phase; (2) the hot thermophilic phase; (3) the cooling phase; and (4) the curing phase.

Compost bacteria combine carbon with oxygen to produce carbon dioxide and energy. Some of the energy is used by the microorganisms for reproduction and growth; the rest is given off as heat. When a pile of organic material begins to compost, mesophilic bacteria reproduce and multiply, raising the temperature of the composting mass up to about 111°F (44°C). This is the first stage of the composting process.

The thermophilic bacteria start to take over in the transition range of 111°F to 125.6°F (44°C–52°C). This begins the second stage of the process, when thermophilic microorganisms become very active and produce a lot of heat. This stage can then continue to about 158°F (70°C) in larger compost piles, although such high temperatures may not be common in smaller backyard compost bins. This heating stage takes place rather quickly and can last a few days, weeks, or many months depending on the amount and nature of the material being composted.

The hot area tends to be localized in the central portion of a compost bin, which is where you should be adding your fresh material. In batch compost, the entire composting mass may become thermophilic at once.

The thermophilic phase wipes out pathogens rather quickly, after which most of the organic material will appear to have been digested, but the coarser organic material will not. This is when the third stage of composting, the cooling phase, takes place. During this phase, the microorganisms that were chased away by the thermophiles migrate back into the compost and get to work digesting the more resistant organic materials. Fungi and macroorganisms such as earthworms and sow bugs help break down the coarser elements into compost.

After the thermophilic stage has been completed, only the readily available nutrients in the organic material have been digested. There's still a lot of food in the pile, and a lot of work to be done by the creatures in the compost. It takes many months to break down some of the more resistant organic materials such as lignin, which comes from wood materials. Like humans, trees evolved with a skin resistant to bacterial attack, and in a compost pile these lignins resist breakdown by thermophiles. However, other organisms, such as fungi, can break down lignin, over time.

The final stage of the composting process is called the curing, aging, or maturing stage, a time period that adds a safety net for pathogen elimination. Many human pathogens have only a limited viability in the soil,

Compost grows plants, which can produce food for humans, thereby providing nourishment and enhancing the health of the individual, the family, and the community.

Compost supports organic gardening where plants
can be grown without the need for harmful chemicals.

and the longer they are subjected to the microbiological competition of the compost pile, the more likely they will be eradicated. Immature or uncured compost produces substances called phytotoxins that are toxic to plants. It can also rob the soil of oxygen and nitrogen and can contain high levels of organic acids. So let your compost reach full maturity before you use it.

The microbes will tell you when they're done. Keep a compost thermometer in your pile and leave it there once you've finished building it. Shove it right into the center of the pile so the dial is down against the surface of the cover material. Don't move it. As the pile shrinks, the dial will appear to move up from the surface, although the surface is really moving down. This shows you the amount of shrinkage that is occurring. Also, the dial will show you the temperature of the pile. Once that temperature has reached ambient (outdoor air temperature), then your pile is likely finished. If in doubt, take a sample of compost from the pile, put it in a small container, and germinate a seed in it, maybe a cucumber, squash, or pumpkin seed. If the compost is immature, the seedling will look unhealthy.

If you keep stirring up your compost, you risk cooling it down prematurely. You may think it's finished after, say, three months, but it would still be composting if you had just left it alone. The author has seen undisturbed compost in cubic meter piles stay above 131°F (55°C) for six months or longer, and larger undisturbed piles for over a year. Let the microbes tell you when they're done. Use a compost thermometer and they'll signal you that way. If the compost temperature is above the outside air temperature, the microbes are still busy.

Compost is normally populated by three general categories of microorganisms: bacteria, actinomycetes, and fungi. Actinomycetes are intermediates between bacteria and fungi because they look like fungi and have similar nutritional preferences and growth habits. They tend to be more commonly found in the later stages of compost and are generally thought to follow the thermophilic bacteria in succession. They, in turn, are followed predominantly by fungi during the last stages of the composting process.

There are at least one hundred thousand known species of fungi; most of them are microscopic. Most fungi cannot grow at 122°F (50°C) because it's too hot, although thermophilic fungi are heat tolerant. Fungi tend to be absent in compost above 140°F (60°C), and actinomycetes above 158°F (70°C). Above 180°F (82°C) biological activity effectively stops.

A teaspoon of native grassland soil can contain six hundred to eight hundred million bacteria comprising ten thousand species, plus perhaps five thousand species of fungi, the mycelia of which could be stretched out for several miles. In the same teaspoon, there may be ten thousand individual protozoa of perhaps a thousand species, plus twenty to thirty different nematodes from as many as one hundred species. Good compost will reinoculate depleted, sanitized, chemicalized soils with a wide variety of beneficial microorganisms.

What Not to Compost

Some compost instructors prohibit certain ingredients in a compost pile. Those "banned" materials include meat, fish, milk, butter, cheese and other dairy products, bones, lard, mayonnaise, oils, peanut butter, salad dressing, sour cream, weeds with seeds, diseased plants, citrus peels, rhubarb leaves, crabgrass, pet manures, bread products, rice, tea bags, and perhaps worst of all — humanure.

The author has composted everything on that list and done so continuously for forty-four years at the time of this writing. All tea bags go into his compost, tags and strings included. He puts buckets of citrus peels in the compost. He composted menstrual pads for years and has recycled at least fifty dead animals through his family-size backyard compost bins. Why would it work for him and not for anyone else? Perhaps the answer is *the compost toilet*.

When compost heats up, much of the organic material is quickly degraded. The materials on the "banned" list may require biological heat and compost management for best results. If composters throw their discards on *top* of their compost piles, that half-eaten pork chop (for example) may sit there looking irresistible to a wandering dog, cat, raccoon, or rat.

Ironically, when the forbidden materials, including toilet material, are combined with other organic ingredients, and inserted *into* the compost

Gardens can include a mix of food crops, ornamentals, and herbs. Here, oregano is being harvested to use in a pickling brine for preserving beans and cucumbers.

pile, thermophilic conditions may prevail. When humanure and the other controversial organic materials are segregated from compost, thermophilic conditions may not occur — a situation that is probably common in many backyard compost piles. The solution is not to segregate materials from the pile, but to add nitrogen and moisture, as are commonly found in manure. Granted, some things do not compost very well. Bones are one of them, but they do no harm in a compost pile.

Don't compost things that microbes don't want to eat, such as ashes (wood or coal), ground minerals, heavy metals, plastics, rubber, glass, and synthetic fertilizers. Having a compost pile is like having a goat in your backyard. Keep it fed with what it likes to eat, and you'll keep it happy!

When the human nutrient cycle remains intact, what would have been waste becomes food, which must be harvested, processed, preserved, prepared, and eaten.

Legal Considerations

Laws and regulations regarding composting vary from place to place. In most of the world, it is unlikely that any laws exist prohibiting or regulating composting on a household level. Waste disposal is often regulated, and it should be. Waste disposal is potentially dangerous to the environment. Sewage disposal and recycling are also regulated in most countries, and they should be, too. Sewage includes a host of hazardous substances from a variety of sources, both domestic and industrial, deposited into a waterborne waste stream. The process of composting is neither disposal of waste nor production of sewage — it is the recycling of organic material.

Gardens provide security. Food is produced, preserved, and available without reliance on outside sources, providing nourishment for the body, the soil, and the soul.

Both backyard composting and farm composting are generally exempt from regulations in the US unless the compost is being sold or removed from the property on which it is made, or the compost operation is large.

Composting is not a wastewater treatment system, does not involve wastewater in pipes, and does not produce or dispose of septage. Composting is not subject to the regulations that govern such systems.

Commercial "dry toilets" produce septage, and even though they may be incorrectly referred to as "composting" toilets, no composting is involved in such toilet systems. Dry toilets that dehydrate and degrade the organic material inside them are regulated in many states. A compost toilet does not degrade organic material; it simply collects it. If the composting occurs on private property and the compost is not being sold, it is most likely not regulated. You should be able to freely compost your organic material if you are not creating a nuisance or giving anybody something legitimate to complain about, such as odors, rats, or liquid leaching out of your pile onto someone's property, all of which are easily avoided.

A mistake the author has seen is when people in the US approach wastewater authorities and ask for permission to make compost. Wastewater professionals are not educated in compost science and will instead attempt to imagine a compost toilet system as being some sort of wastewater treatment system, which it is not. Involving authorities who have no, or limited, understanding of composting can be opening a bag of worms that may not be easily closed.

Modern governments typically enact legislation that encourages the development of resource recovery as a means of reducing waste and conserving resources. Since the composting of toilet material involves recovering a resource, requires no disposal of anything, and is ecologically friendly, it is a commendable sanitation option.

It's not hard to do a good job of making compost. The most likely problem you could have is an odor problem, and that would simply be due to not keeping your compost adequately covered with clean cover material. If you keep it covered, it does not give off offensive odors. It's that simple. Human excretions smell bad so people will be naturally compelled to cover them with something. That makes sense when you think that thermophilic bacteria are already in the feces waiting for the manure to be piled into a compost bin with plant material, so they can get to work.

Admittedly, many humans suffer from arrested development when it comes to understanding human excretions. What we put *into* our bodies, food for example, is celebrated as an art and a science. Ironically, what comes *out* of our bodies is ignored, neglected, and avoided. We are held captive to a nineteenth century attitude that our excretions are waste materials choking with disease organisms, when in fact our excretions are recyclable organic materials teeming with beneficial microorganisms.

People who are trying to advance sanitation, and especially the practical applications of composting, sometimes face resistance from regulatory personnel, primarily in water toilet cultures. This same pushback is rare in areas where water toilets do not exist. On the contrary, government authorities in developing nations tend to be keenly interested in compost toilets and want to expand their use to reduce or eliminate pit latrines, open defecation, and water pollution, while enhancing soil fertility.

One obstacle to expanding compost sanitation in developing nations is lack of financial support for training and logistics. The author had compost toilets constructed in Mongolia at a cost of $12 US each, including labor. GiveLove.org built compost toilets in Haiti for $25, and in Africa for $50 each, which included training and construction of a compost bin for each family. In many areas of the world people live on $2 or less a day. They still need a toilet. The richest 1 percent of humans now have as much wealth as half of humanity. It is an unfortunate reflection on the human species when a handful of people can hoard so much wealth while billions of people can't even afford a simple toilet.

Chapter Nine

HEALTH, SAFETY, AND THERMOPHILES

Much of the pathogen information in this section is adapted from *Appropriate Technology for Water Supply and Sanitation*, by Feachem et al., World Bank, 1980.

A discussion of compost toilets is not complete without a review of the health and safety implications. Human excrement and sewage have been irresponsibly discarded for centuries. This has created epic health catastrophes, such as widespread cholera from polluted drinking water, epidemics of intestinal parasites from polluted soil and food; even the bubonic plague, carried by fleas on rats, was likely enhanced by the festering garbage piles and sewage dumps on which the rats dined.

The responsible recycling of toilet material and organic "garbage" by composting is a new human concept. Only in recent history have we grasped the awareness of microorganisms and their association with disease. Even more recently have we gained awareness of microorganisms and their role in recycling organic materials. Composting, and its ability to eliminate human disease organisms (pathogens), is a subject that is virtually unknown among many human societies.

Feces can contain disease organisms that can contaminate the environment and infect people when they are discarded as a waste material and pollutant. Even a healthy person apparently free of disease can pass potentially dangerous pathogens through his or her feces, simply by being a carrier. The World Health Organization estimates that 80 percent of all diseases are related to inadequate sanitation and polluted water, and that half of the world's hospital beds are occupied by patients who suffer from water-related diseases. When toilet material is not composted, and sewage is dispersed into the environment, various diseases and parasites can infect the population living in the contaminated areas.

On the other hand, disease organisms are not spread by properly prepared compost. There is no reason to believe that the manure of a human being is dangerous unless it is allowed to accumulate in the environment, pollute soil or water, or breed flies and rats, all of which are the results of

The world is filled with elderly and infirm persons who don't have toilets. The 80-year-old man above lives alone in a very humble abode (opposite page) without electricity or running water, in central Nicaragua. His new compost toilet was placed next to his bed, giving him the first indoor toilet of his lifetime. A compost bin was constructed for him behind his dwelling, and rice hulls were provided for cover material.

negligence, or ignorance. Collecting and composting toilet material can provide hygienically safe, ecological sanitation without the use of dangerous chemicals, prohibitive costs, or a high level of technology and energy consumption.

Compost kills human disease organisms. This is important, well-established science. It's what makes composting such a valuable practice. A combination of factors inhibits pathogens in compost, including:

- Competition for food from compost microorganisms
- Inhibition and antagonism by compost microorganisms
- Consumption by compost organisms
- Biological heat generated by compost microorganisms
- Antibiotics produced by compost microorganisms.

For example, when pathogens were grown in an incubator *without compost* at 122°F (50°C) and separately *with compost* at 122°F, they died in the compost after only seven days, but lived in the incubator for seventeen days. This indicated that it is more than just temperature that determines the fate of pathogenic bacteria. The other factors listed above undoubtedly affect the viability of nonindigenous microorganisms, such as human pathogens, in a compost environment.

Those factors benefit from a diverse microbial population, a diversity which is best achieved by temperatures below 140°F (60°C). One researcher stated that significant reductions in pathogen numbers have been observed in compost piles which have not exceeded 104°F (40°C).

When Westerberg and Wiley composted sewage sludge inoculated with polio virus, *Salmonella*, roundworm eggs, and *Candida albicans*, they found that after forty-three hours of composting, "no viable indicator organisms could be detected," with the polio virus being inactivated in the first hour. They concluded that a compost temperature of 140°F to 158°F (60°C-70°C) maintained for three days would kill all of the pathogens.[1] This has been confirmed by many other researchers, including Gotaas, who indicates that pathogenic bacteria are unable to survive compost temperatures of 131°F to 140°F (55°C–60°C) for more than thirty minutes to one hour.[2]

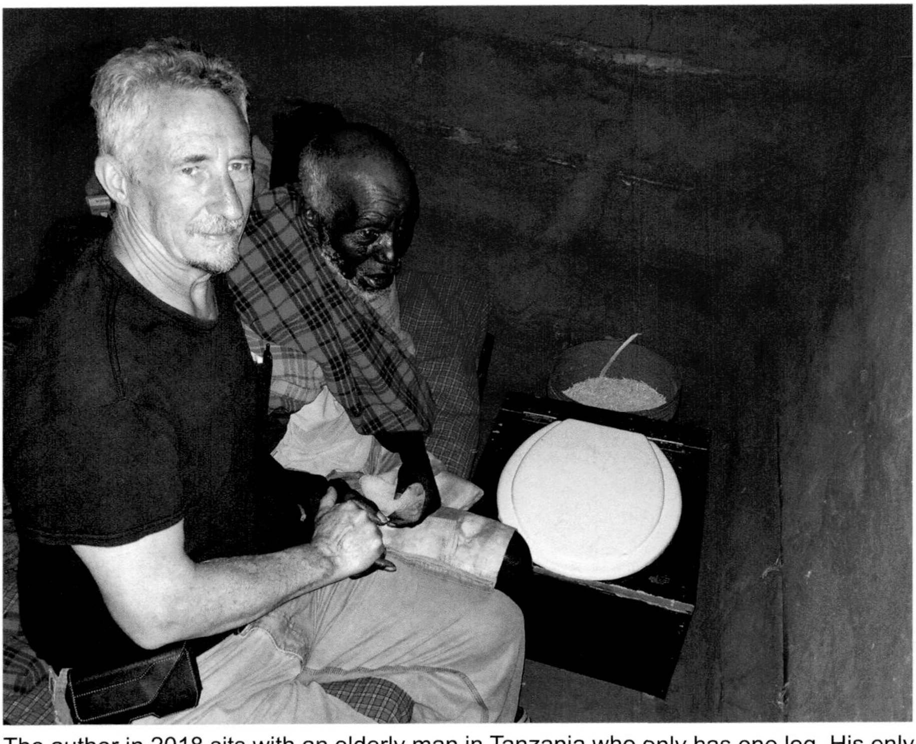

The author in 2018 sits with an elderly man in Tanzania who only has one leg. His only toilet option was squatting over a hole in a distant pit latrine until he had a compost toilet installed next to his bed. His son-in-law and teenage granddaughter manage the composting process for him. Reluctance by NGOs, government agencies, and health authorities to accept composting as a viable sanitation option can deprive many needy people of the comfort and convenience of a secure, low-cost, odorless indoor toilet.

References from page 134: [1] Wiley B. Beauford and Stephen C. Westerberg. "Survival of Human Pathogens in Composted Sewage." *Applied Microbiology* [01 Dec 1969, 18(6):994-1001]. [2] Gotaas, Harold B. "Composting — Sanitary Disposal and Reclamation of Organic Wastes." Geneva: World Health Organization, Monograph Series Number 31. p. 20. [Read more in *The Humanure Handbook*, 4th edition.]

People may believe that it's only the heat of the compost pile that destroys pathogens, so they want their compost to become as hot as possible. This can be a mistake. In fact, compost can become too hot, and when it does, it destroys the biodiversity of the microbial community and may actually constitute a barrier to effective sanitization under certain circumstances. Perhaps only one species (i.e., *Bacillus stearothermophilus,* otherwise known as *Geobacillus*) may dominate the compost pile during periods of excessive heat, thereby driving out or outright killing the other inhabitants of the compost, which include fungi and actinomycetes as well as the bigger organisms that you can actually see. Control of excessive heat, however, is unlikely to be a concern for the small-scale composter, because smaller masses of organic material typically do not develop temperatures as high as larger masses.

THERMOPHILES

Bacteria are generally divided into three classes based on the temperatures at which they best thrive. The low-temperature bacteria are the psychrophiles, whose optimum temperature is 59°F (15°C) or lower. The mesophiles live at medium temperatures between 68° and 113°F (20° and 45°C). Thermophiles thrive above 113°F, and some live at, or even above, the boiling point of water.

Perhaps the most mysterious and impressive bacteria are the thermophiles, or heat lovers. If you want to conduct an experiment, grind fresh tree branches with a chipper-shredder machine, pile the tree particles in a heap and stick a compost thermometer in it. You will find that within about seventy-two hours, the internal temperature of the ground-up tree will be 120° to 130°F (49°–54°C), if the pile is big enough. The heat is from thermophilic bacteria, microorganisms that seem to live everywhere. These bacteria prefer a hot environment and will even *create* one when given a chance. But why are they in trees? Why do they also exist in cool environments? Where do they come from?

Geobacillus, formerly named *Bacillus stearothermophilus,* is an aerobic, rod-shaped bacterium that forms spores. Its temperature range for growth

TEMPERATURE CONVERSIONS

F	C
-40	-40
-30	-34.44
-20	-28.88
-10	-23.33
0	-17.77
5	-15.00
10	-12.22
15	-9.44
20	-6.66
25	-3.88
30	-1.11
35	1.66
40	4.44
45	7.22
50	10.00
55	12.77
60	15.55
65	18.33
70	21.11
75	23.88
80	26.66
85	29.44
90	32.22
95	35.00
98.6	36.99
100	37.77
105	40.55
110	43.33
115	46.11
120	48.88
125	51.66
130	54.44
135	57.22
140	60.00
145	62.77
150	65.55
155	68.33
160	71.11
165	73.88

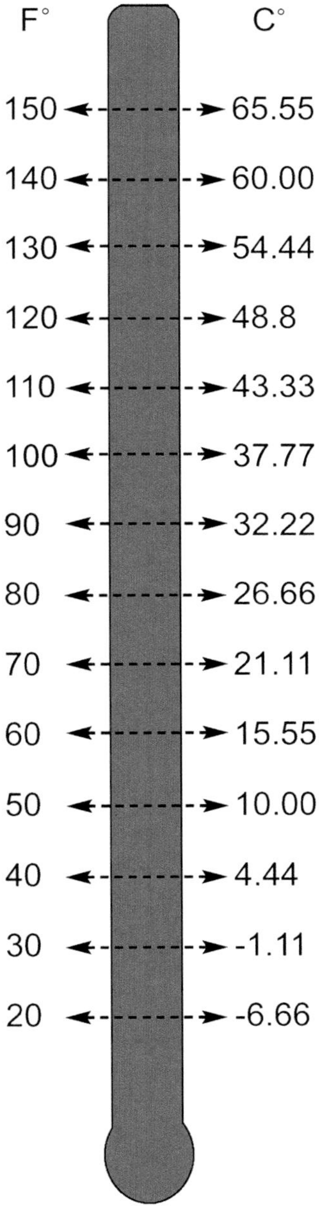

C	F
0	32.00°
5	41.00°
10	50.00°
15	59.00°
20	68.00°
25	77.00°
30	86.00°
35	95.00°
40	104.00°
45	113.00°
50	122.00°
55	131.00°
60	140.00°
65	149.00°
70	158.00°
75	167.00°
80	176.00°
85	185.00°
90	194.00°
95	203.00°
100	212.00°

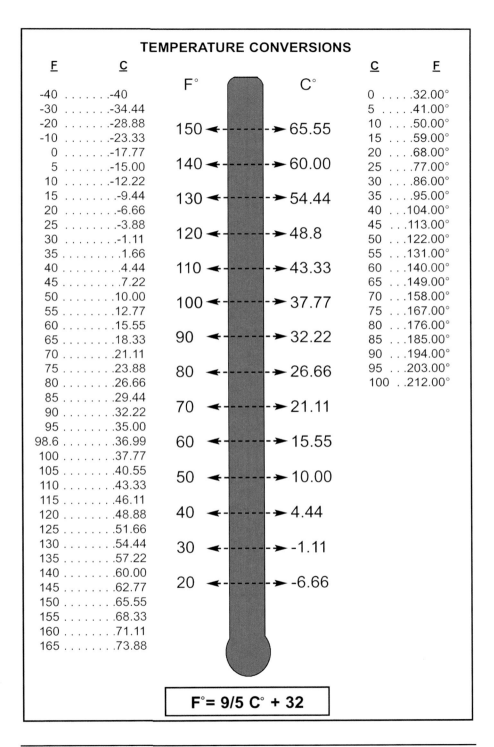

$$F^{\circ} = 9/5\, C^{\circ} + 32$$

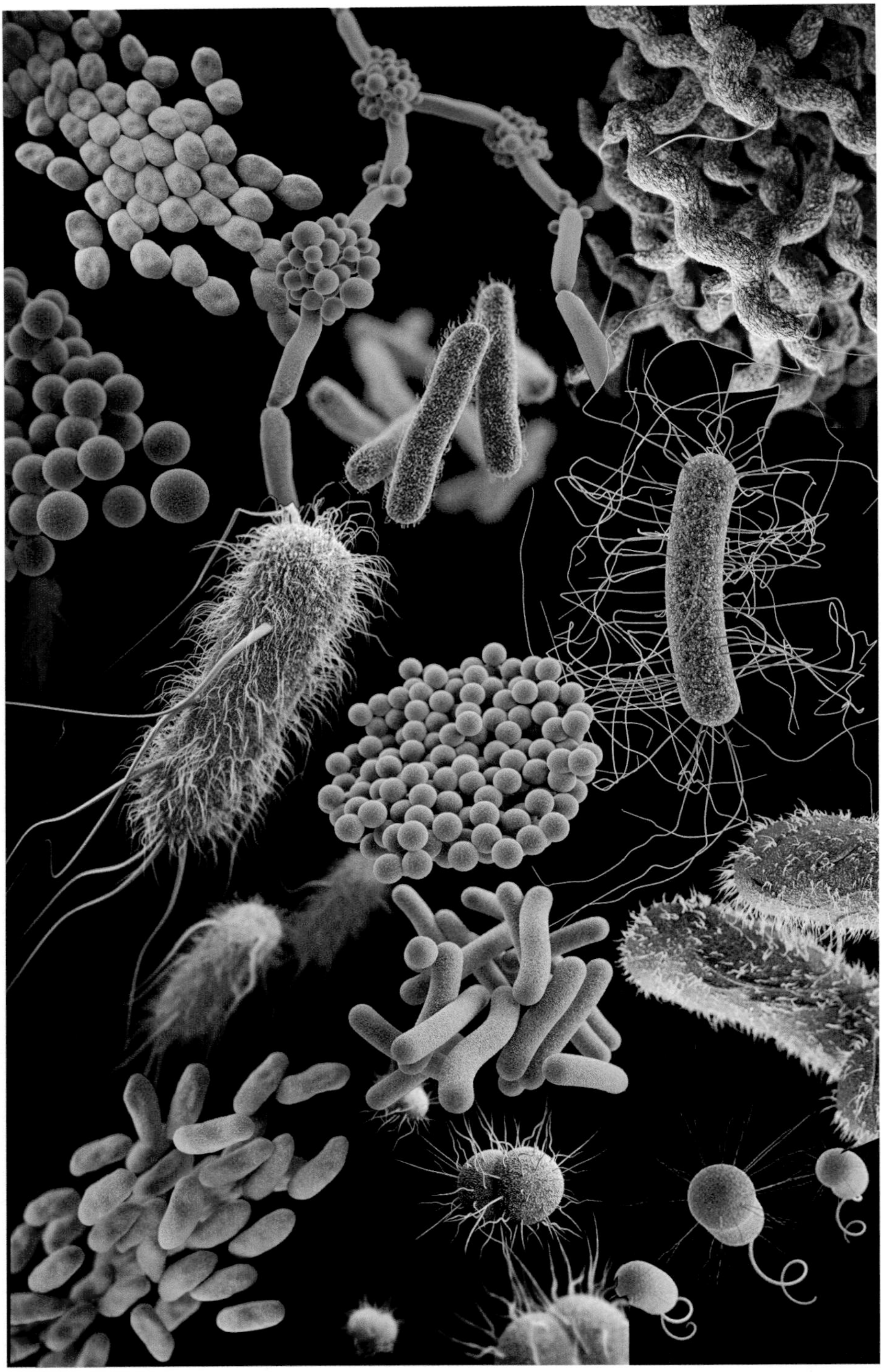

can be as low as 95° or as high as 176°F (35° or 80°C), but temperatures between about 113° and 158°F (45° and 70°C) are normal. Despite this very hot optimum temperature range, these mysterious thermophilic bacteria can be found all over the Earth, a planet with an average surface temperature between 44° and 50°F (7° and 10°C).

Thermophilic bacteria have been found on all seven continents, in the Pacific Ocean, in the Mediterranean Sea, in the Bolivian Andes at an altitude of 12,000 feet, and even in the upper troposphere 6 miles high. They've been found in oil wells 7,000 feet below the ground, in gold mines 10,000 feet underground, and in the ocean over 6 miles below sea level. These are bacteria that love to eat human excrement, discarded organic material, and dead animals, yet their favorite hangouts are hot springs, geothermal soils, hot underground oil fields, natural gas wells, and hydrothermal vents. And compost piles.

Scientists wonder how thermophilic bacteria can exist in large numbers in cool environments where they can't grow. They speculate that the Earth's population of thermophiles is "enormous." The answer seems to lie in the ability of thermophilic bacteria to form spores. When they don't have conditions favorable for growth (i.e., high temperatures), they form "endospores," a life cycle condition that allows for their long-term survival. One theory suggests that the thermophiles were among the first living things on this planet, developing and evolving during the primordial birthing of the Earth when surface temperatures were quite hot. They have thus been called the Universal Ancestor, estimated at 3.6 billion years old. Thermophiles could therefore be the common ancestral organism of all life-forms on our planet.

Thermophilic bacteria have evolved to decompose organic material, almost like the Earth's janitors, or maybe Mother Earth's invisible helpers. They work in partnership with mesophilic bacteria, which raise the temperature of an organic mass high enough for thermophilic growth to be sparked. This is like a microbial tag team — mesophiles begin the decomposition; this raises the temperature enough to waken the thermophilic spores; the work is then handed off to the thermophiles, who take over and

Opposite: Electron micrographs of bacteria from the US Centers for Disease Control.

work themselves into a fever, consuming the organic material, be what it may (turds, garbage, dead animals), and converting it back into, well, Mother Earth. In the process, if there happen to be human pathogens lurking in the organic material, they're no match for the thermophiles.

Mother Earth has some interesting tricks up her sleeves. On every square meter of her surface, she is releasing fifty to over two hundred bacteria per second. These can be uplifted by the wind, where they can remain aloft for two to fifteen days before settling back to Earth. That's probably

Compost toilet training in the Zombo district of northern Uganda, Africa (an International Medical Outreach project).
Photo by Samuel Autran Dourado.

why some thermophiles have been found way up in the troposphere. Add massive wind events such as desert dust storms, which can move a billion tons of soil a year, and bacteria can cross the entire Atlantic Ocean in just three to five days and the Pacific Ocean in a week to ten days.

Thermophilic spores are hardened to survive. They're resistant to desiccation, ultraviolet light, and extreme temperatures, conditions that will kill most other bacteria. So Mother Earth scatters them around the globe as microscopic spores, a form that is unbelievably durable and long-lasting. They settle back to the Earth and wait patiently, even a billion years if necessary, to be sparked back to life. They're Mother Nature's servants, patiently waiting to serve *us*.

It seems miraculous that Mother Nature provides us with a simple, inexpensive, nonchemical, nonpharmacological, nontechnical, biological tool for the elimination of disease organisms — a tool accessible to almost anyone, anywhere, if they only knew it existed. But it does exist, and it is perhaps this characteristic of compost that most gives it its unique value and sets it apart from other forms of organics recycling. This is also why professionals in the compost industry don't want noncomposted degraded organic material incorrectly referred to as "compost."

Some thermophilic actinomycetes, as well as mesophilic bacteria, produce antibiotics that display considerable potency toward other bacteria. Up to one-half of thermophilic strains can produce antimicrobial compounds, some of which have been shown to be effective against *E. coli* and *Salmonella*. The production of antibiotics by compost microorganisms theoretically assists in the destruction of human pathogens that may exist in organic material.

Even if every speck of the composting material is not subjected to the high internal temperatures of the compost pile, the composting process nevertheless contributes immensely toward the creation of sanitary organic material. Or in the words of one group of composting professionals, the high temperatures achieved during composting, assisted by the competition and antagonism among the microorganisms (i.e., biodiversity), considerably reduce the number of plant and animal pathogens. While some resistant pathogenic organisms may survive, and others may persist in cooler sections of the pile, the disease risk is, nevertheless, greatly reduced.

If a small-scale or community composter has any doubt or concern about the existence of pathogenic organisms in his or her finished compost, s/he can use the compost for horticultural purposes rather than for food production. Furthermore, lingering pathogens continue to die after the compost has been applied to the soil, which is not surprising since most human pathogens prefer the warm and moist environment of the human body. As World Bank researchers put it, even pathogens remaining in compost seem to disappear rapidly in the soil.

Compost can also be tested for pathogens by testing labs. The idea that compost must be sterile is incorrect. It must be sanitary, which means it must have a greatly weakened, reduced, or destroyed pathogen population.

PATHOGENS

The pathogens, or disease organisms, that can exist in toilet material can be divided into four general categories: viruses, bacteria, protozoa, and worms (helminths).

VIRUSES: First discovered in the 1890s by a Russian scientist, viruses are among the simplest and smallest biological entities. Many scientists don't even consider them to be organisms. They are much smaller and simpler than bacteria — the simplest form may consist of only an RNA molecule. By definition, a virus is an entity that contains the information necessary for its own replication but does not possess the physical elements for such replication. To reproduce, viruses must infect a host cell, which is reprogrammed by the virus. Viruses cannot reproduce outside their host.

There are more than 140 types of viruses worldwide that can be passed through human feces, including polioviruses, coxsackieviruses (causing meningitis and myocarditis), echoviruses (causing meningitis and enteritis), reovirus (causing enteritis), adenovirus (causing respiratory illness), and infectious hepatitis (causing jaundice). During periods of infection, one hundred million to one trillion viruses can be excreted with each gram of fecal material.

BACTERIA: Of the pathogenic bacteria, the genus *Salmonella* is significant because it contains species causing typhoid fever, paratyphoid, and gastrointestinal disturbances. Another genus of bacteria, *Shigella*, causes dysentery. Myobacteria cause tuberculosis.

PROTOZOA: The pathogenic protozoa include *Entamoeba histolytica* (causing amoebic dysentery), and members of the Hartmannella-Naegleria group (causing meningo-encephalitis). The cyst stage in the life cycle of protozoa is the primary means of dissemination as the amoeba die quickly once outside the human body. Cysts must be kept moist to remain viable for any extended period.

PARASITIC WORMS: A number of parasitic worms pass their eggs in feces, including hookworms, roundworms (*Ascaris*), and whipworms. These eggs tend to be resistant to environmental conditions because of their thick outer covering, and they are extremely resistant to the sludge digestion process common in wastewater treatment plants. Roundworms coevolved as parasites of humans by taking advantage of the human habit of defecating on soil. Since roundworms live in human intestines but require a period in the soil for their development, their species is perpetuated by our excretory habits. If we humans never allow our excrement to come in contact with soil, and if we instead compost it, the parasitic species known as *Ascaris lumbricoides*, a parasite that has plagued humans for perhaps hundreds of thousands of years, would soon become extinct.

INDICATOR PATHOGENS: Indicator pathogens are those whose detection in soil or water serves as evidence of fecal contamination. The *Ascaris lumbricoides* (roundworm) egg is persistent and can serve as an indicator for the presence of pathogenic helminths in the environment. The reported viability of roundworm eggs in soil ranges from a couple of weeks under sunny, sandy conditions to two and a half years, four years, five and a half years, or even ten years in soil, depending on the source of the information. Consequently, the eggs of the roundworm seem to be the best indicator for determining if parasitic worm pathogens are present in soil or compost.

Handwashing station at a school in northern Uganda. The compost bin and toilet stalls are visible in the background. Prior to the compost toilets, the students used pit latrines.

Ascaris eggs develop at temperatures between 60° and 95°F (15.5° and 35°C), but the eggs disintegrate at temperatures above 100.40°F (38°C). The temperatures generated during composting can easily exceed levels necessary to destroy roundworm eggs.

There are indicators other than roundworm eggs that can be used to determine fecal contamination of water, soil, or compost. Indicator *bacteria* include fecal coliforms, which reproduce in the intestinal systems of warm-blooded animals. If one wants to test a water supply for fecal contamination, then one looks for fecal coliforms, usually *Escherichia coli*, which is among the most abundant intestinal bacteria in humans; over two hundred specific types exist. Although some of them can cause disease, most are harmless, or even beneficial. The absence of *E. coli* in water indicates that the water is free from fecal contamination.

Fecal coliforms do not multiply outside the intestines of warm-blooded animals; therefore, their presence in water is unlikely unless there is fecal pollution. Since fecal coliforms survive for a shorter time in natural waters than the coliform group as a whole, their presence indicates relatively recent pollution. Almost all of our natural waters have a presence of fecal coliforms, since all warm-blooded animals excrete them. Bacterial analyses of drinking water supplies are routinely provided for a small fee by agricultural supply firms, water treatment companies, or private labs.

PRIONS: According to the US Environmental Protection Agency, it has been found that bovine spongiform encephalopathy (BSE), or mad cow disease, is caused by a prion protein. The pathway for transmission is by eating tissue from infected animals. There has been no evidence that the BSE prion protein is shed in feces or urine. The primary route for infection, the use of animal carcasses in animal feed, is banned in the US. There should be no risk of BSE exposure from compost.

HIV: Is there any risk of human immunodeficiency virus (HIV) infection from compost? The EPA reports that the HIV virus is contracted through contact with blood or other body fluids of an infected individual. According to the Danish publication "Sustainable Urban Renewal and

Wastewater Treatment: Occurrence and Survival of Viruses in Composted Human Feces," No. 32, 2003, *Contamination with urine or blood may lead to the occurrence in feces of . . . human immunodeficiency virus (HIV) and the hepatitis B virus (HBV). However, such viruses are not of particular relevance in relation to composting of human feces, due to their sporadic occurrence in human feces, their poor survival in the environment, and their non-intestinal route of transmission. These factors minimize the risks of human exposure to these viruses via the production and usage of composted human feces.*

PINWORMS: Pinworms (*Enterobius vermicularis*) are common among young children. These unpleasant parasites are spread from human to human by direct contact. The pinworm life cycle does not include a stage in soil, compost, or manure. Pinworms lay microscopic eggs at the anus of a human being, its only known host. This causes itching at the anus which is the primary symptom of pinworm infection. The eggs can be picked up almost anywhere. Once in the human digestive system, they develop into the tiny worms. Infection is spread by the hand-to-mouth transmission of eggs resulting from scratching the anus, as well as from breathing airborne eggs. In about one-third of infected children, eggs may be found under the fingernails. In 95 percent of infected persons, pinworm eggs aren't found

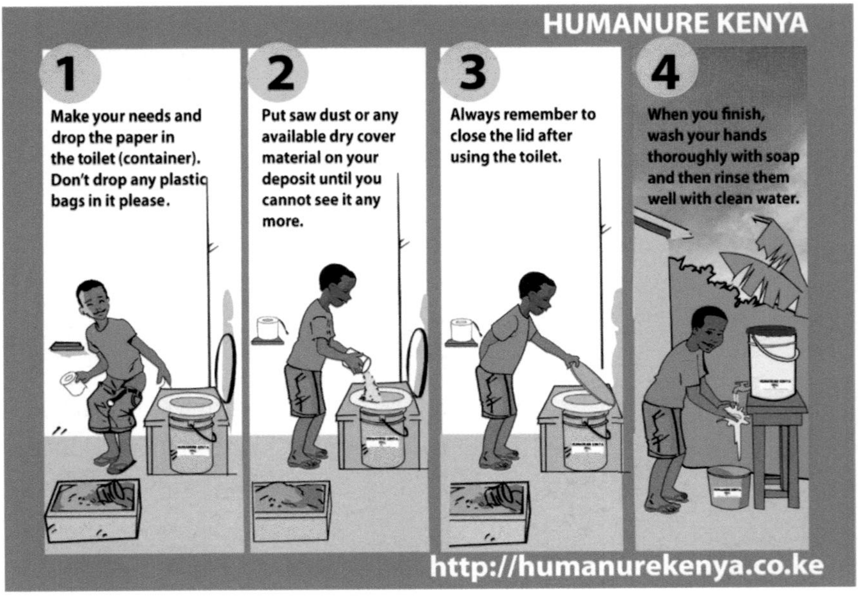

in the feces. Transmission of eggs to feces and to soil is not part of the pinworm life cycle, which is one reason the eggs aren't likely to end up in feces or compost. Even if they do, they quickly die outside the human host.

HOOKWORMS: Hookworm species in humans include *Necator americanus*, *Ancylostoma duodenale*, *A. braziliense*, *A. caninum*, and *A. ceylanicum*. These small worms are about a centimeter long (less than half an inch); humans are almost the exclusive host of *A. duodenale* and *N. americanus*. A hookworm of cats and dogs, *A. caninum*, is an extremely rare intestinal parasite of humans. The eggs are passed in the feces and mature into larvae outside the human host under favorable conditions. The larvae attach themselves to the bottom of your foot when they're stepped on, then enter your body through pores, hair follicles, or even unbroken skin. They tend to migrate to the upper small intestine where they suck their host's blood. Within five or six weeks, they'll mature enough to produce up to twenty thousand eggs per day. Don't walk barefoot around pit latrines! Both the biological temperatures of composting and the freezing temperatures of winter will kill the eggs and larvae. Dehydration is also destructive.

WHIPWORMS: Whipworms (*Trichuris trichiura*) are usually found in humans but may also be found in monkeys or hogs. They're usually under 2 inches long; the female can produce three thousand to ten thousand eggs per day. Larval development occurs outside the host. In a favorable environment (warm, moist, shaded soil), first-stage larvae are produced from eggs in three weeks. The life span of the worm is usually considered to be four to six years. Persons handling soil that has been defecated on by an infected person risk infection by hand-to-mouth transmission of the eggs. Light infections may not show any symptoms. Heavy infections can result in anemia and death. A stool examination will determine if there is an infection. Cold winter temperatures of 18° to 10°F (–8° to –12°C) are fatal to the eggs, as are the high temperatures of composting.

ROUNDWORMS: Roundworms (*Ascaris lumbricoides*) are fairly large worms, 10 inches in length (25 cm), that parasitize the human host by eating semi-digested food in the small intestine. The females can lay two hundred thousand eggs per day for a lifetime total of roughly twenty-six million. Larvae develop from the eggs in soil under favorable conditions (70°–86°F [21° – 30°C]). Above 99°F (37°C), they cannot fully develop.

Eggs are destroyed by direct sunlight within fifteen hours and are killed by temperatures above 104°F (40°C), dying within an hour at 122°F (50°C). The eggs are resistant to freezing, chemical disinfectants, and other strong chemicals, but composting will kill them.

Roundworms, like hookworms and whipworms, are spread by fecal

One common concern is that the compost isn't getting hot around the edges. The video screen grab shows a 20-inch compost thermometer pulled out of a compost bin in Haiti and stuck off at the edge out of the way while toilet material was being added in the center of the bin. Even though the thermometer was barely penetrating the cover material, at the edge of the pile, the temperature reading was 130°F (54.4°C) just below the surface of the material. This illustrates the importance of keeping a cover material on and around the compost pile. The cover material not only insulates the pile, but it also protects the pile from excess rainfall as well as from cold outdoor temperatures.

contamination of soil. Much of this contamination is caused and spread by children who defecate outdoors on the ground. One way to eradicate fecal pathogens is to conscientiously compost all fecal material. Therefore, it is very important when composting humanure to be certain that all children use a toilet facility and do not defecate on the soil. When changing soiled diapers, scrape the fecal material into a compost toilet with toilet paper, leaves, or another biodegradable material. And don't forget to wash your hands after feeding your compost pile and before feeding yourself!

Composting

There is no proven, natural, low-tech, beneficial method for destroying human pathogens in organic material that is as successful and accessible to the average human as composting. But what happens when the compost is not well managed? How dangerous is the undertaking when those involved don't try to ensure that the compost maintains adequate temperatures? In fact, this is usually what happens in most owner-built and commercial dry toilets. Composting does not occur in dry toilets because the correct blend of ingredients or the environment needed for such microbial activity does not exist. In the case of most commercial dry toilets, composting is not even intended. Instead, the toilets are designed to be dehydrators rather than composters.

What happens to ignored and neglected outdoor compost piles? After a couple of years, they turn into a pile of soil, and if left entirely alone, they simply become covered with green vegetation, eventually blending back into the earth.

What about toilet material from a diseased population? Such a population would, for example, be the residents of a hospital in an underdeveloped country, or residents in a community where certain diseases or parasites are endemic. In that situation, the composter must make every effort necessary to ensure well-managed composting, adequate retention time, and adequate pathogen elimination. Dedicated gloves, boots, tools, even coveralls and a dust mask would be recommended.

According to Dr. T. Gibson, head of the Department of Agricultural

Biology at the Edinburgh and East of Scotland College of Agriculture, "All the evidence shows that a few hours at 120° Fahrenheit would eliminate [pathogenic microorganisms] completely. There should be a wide margin of safety if that temperature were maintained for 24 hours."

Complete pathogen destruction is guaranteed by arriving at a temperature of 143.6°F (62°C) for one hour, 122°F (50°C) for one day, 114.8°F (46°C) for one week, or 109.4°F (43°C) for one month. It appears that no excreted pathogen can survive a temperature of 149°F (65°C) for more than a few minutes. A compost pile can rise to a temperature of 131°F (55°C) or above or maintain a temperature hot enough for a long enough period of time to destroy human pathogens beyond a detectable level.

The United States Environmental Protection Agency (EPA) publishes requirements for the safe reuse of sewage sludge (biosolids) and domestic septage, such as from a dry toilet. The EPA states, "Composting creates a marketable end product that is easy to handle, store, and use. It is usually a 'Class A' material without detectable levels of pathogens that can be applied to gardens, food and feed crops, and rangelands. Biosolids compost

One of the author's backyard compost piles reached a high of about 156F (about 69C) in 2020. This pile had only toilet material, rotted sawdust, and dead animals in it.

is safe to use and generally has a high degree of acceptability by the public. Thus, it competes well with other bulk and bagged products available to homeowners, landscapers, farmers, and ranchers." EPA requirements for "Class A" sewage sludge compost include the following time/temperature requirements:

(1) Aerated static pile or in-vessel: 131°F (55°C) for at least 3 days.

(2) Open windrow: 131°F (55°C) for at least 15 days with 5 turns.

Temperature and Time

Two primary factors lead to the death of pathogens in compost. The first is temperature. A compost pile that is properly managed will destroy pathogens with the heat and biological activity it generates.

The second factor is time. The lower the temperature of the compost, the longer the retention time needed for the destruction of pathogens. Given enough time, the wide biodiversity of microorganisms in the compost will destroy pathogens by the antagonism, competition, consumption, and antibiotic inhibitors provided by the beneficial microorganisms.

One need not strive for extremely high temperatures in a compost pile to feel confident about the destruction of pathogens. It may be more realistic to maintain lower temperatures in a compost pile for longer periods of time, such as 122°F (50°C) for twenty-four hours, or 115°F (46°C) for a week. According to one source, "All fecal [pathogenic] microorganisms, including enteric viruses and roundworm eggs, will die if the temperature exceeds 114.8°F (46°C) for one week."

Compost the toilet material, then allow the compost to sit, undisturbed, for a lengthy retention time to allow the compost to thoroughly age or cure. The biodiversity of the compost microorganisms will aid in the elimination of pathogens as the compost ages. If one wants to be particularly cautious, such as when managing compost toilets for a hospital, one could allow the compost to sit for two years after the pile has been completed, instead of the one year that is typically adequate.

THERMAL DEATH POINTS FOR COMMON PARASITES AND PATHOGENS

PATHOGEN	THERMAL DEATH
Ascaris lumbricoides eggs	Within 1 hour at temps over 50°C
Brucella abortus or *B. suis*	Within 1 hour at 55°C
Corynebacterium diptheriae	Within 45 minutes at 55°C
Entamoeba histolytica cysts	Within a few minutes at 45°C
Escherichia coli	One hr at 55°C or 15-20 min. at 60°C
Micrococcus pyogenes var. *aureus*	Within 10 minutes at 50°C
Mycobacterium tuberculosis var. *hominis*	Within 15 to 20 minutes at 66°C
Necator americanus	Within 50 minutes at 45°C
Salmonella spp.	Within 1 hr at 55C; 15-20 min. at 60°C
Salmonella typhosa	No growth past 46C; death in 30 min. 55C
Shigella spp.	Within one hour at 55°C
Streptococcus pyogenes	Within 10 minutes at 54°C
Taenia saginata	Within a few minutes at 55°C
Trichinella spiralis larvae	Quickly killed at 55°C

Source: Gotaas, Harold B. (1956). Composting - Sanitary Disposal and Reclamation of Organic Wastes . p.81. World Health Organization, Monograph Series Number 31. Geneva.

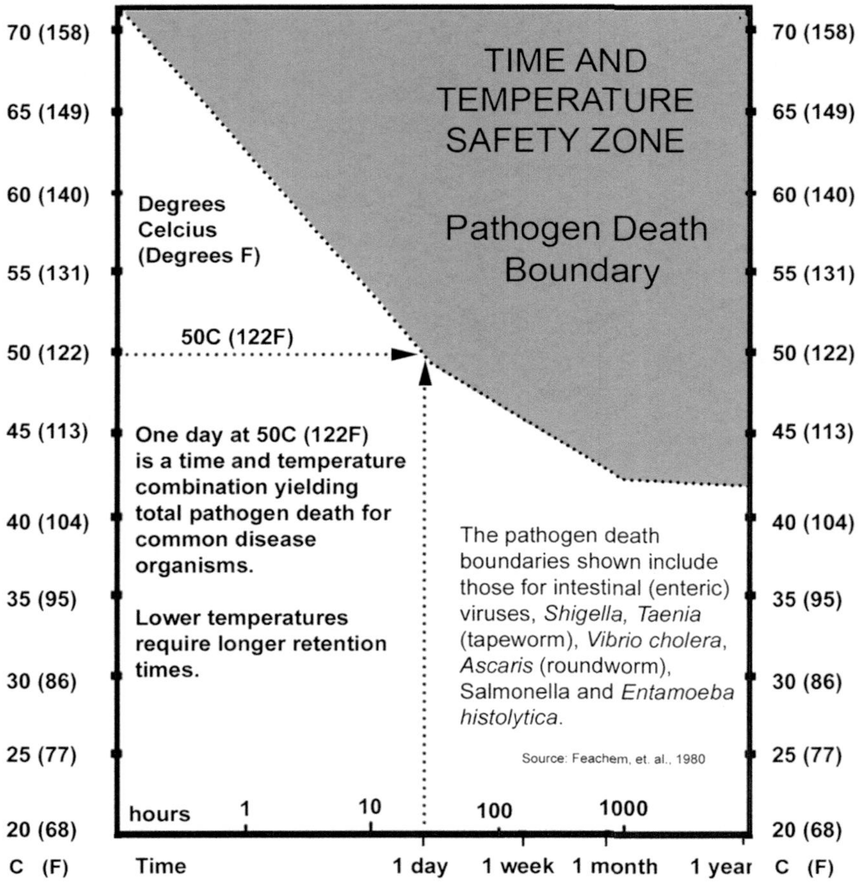

In the words of Feachem, "The effectiveness of excreta treatment methods depends very much on their time-temperature characteristics. The effective processes are those that either make the excreta warm (55°C/131°F), hold it for a long time (one year), or feature some effective combination of time and temperature." The US EPA requires three days at 131°F (55°C) for pathogen elimination in a static compost pile. The author's compost piles in Haiti, California, and elsewhere maintain temperatures above 131°F for many months. These are unturned piles, insulated on top and around the sides with cover material such as straw or sugarcane bagasse. The temperatures are quite uniform throughout the piles, even maintaining pathogen-destroying temperatures right up to the edges.

Monitoring Compost Temperature

Since 1993 the author has monitored his compost temperatures continuously, year-round. The compost typically reaches 120°F (about 50°C) or above, at a depth of 20 inches, in early spring and stays there all summer and fall, as long as the pile is continuing to be fed. In the winter, the temperature can drop, but the compost piles have not frozen since 1997, despite temperatures well below zero F. In fact, the compost's thermophilic bacteria seem to be adapting to the cold winters of Pennsylvania, and it is not uncommon for the compost to read temperatures over 100°F in winter, even when the ambient air temperature is in the single digits (way below zero Celsius). However, the compost piles must be provided with food, or their temperatures will drop.

The maximum temperature the author has recorded in his backyard compost is about 156° F (69°C), but more typical temperatures range from 110°F (43°C) to 130°F (54°C). For some reason, the compost seems to stay around 120°F most of the summer months at a depth of 20 inches. The author keeps his compost thermometers right in the top center of the piles where new material is added. Larger piles, such as for school populations or centralized compost operations in villages, can achieve higher temperatures and hold their temperatures for longer periods of time, simply due to the larger mass.

Large compost bins can be monitored by checking the temperatures at various locations around the bin, and at various depths. Larger bins may require a larger thermometer, perhaps 36 inches or 48 inches long. Rapid response digital thermometers are available, allowing for a quicker reading.

Compost made with toilet material requires approximately a year's undisturbed retention time after the pile is fully built, give or take a month or two, depending on such factors as location, feedstocks, size of the pile, and outdoor ambient temperatures. When the process is preceded by a thermophilic phase, it would be challenging to come up with a more effective, more Earth-friendly, simpler, low-cost system for pathogen elimination, assuming pathogens are there in the first place.

Chapter Ten

PHARMACEUTICALS AND HEAVY METALS

[Note: This subject matter is addressed in greater depth in the fourth edition of *The Humanure Handbook* (Jenkins, 2019), where all reference citations are included. The book can be read free online at HumanureHandbook.com (look for a "read free" link). Each chapter can be downloaded in PDF format, including all reference citations.]

Can people on prescription drugs use a compost toilet? What about pharmaceuticals in compost? Does composting break down the drugs? Are some pharmaceuticals worse than others? Do plants take up pharmaceuticals? What about heavy metals in compost?

To find answers to these questions, research scientists must introduce antibiotics and pharmaceuticals into composting environments, then monitor and record the presence of the pharmaceuticals. Unfortunately, there is not a lot of research on human excrement composting yet. On the other hand, *sewage sludge* compost and *animal manure* composts are available, so researchers look at these materials to get an idea of what composting does to pharmaceuticals and other chemicals that humans and other animals excrete. Our bodies metabolize only a fraction of the drugs we ingest; most of the remainder is excreted into toilets.

In one study, soil contaminated with gout medicine and methaqualone (a sedative) was composted. Results showed that the most effective removal occurred at 77° F (25°C), although the gout medicine removal in the thermophilic stage ranged from 75 to 100 percent. Composting removed the contaminants to an "acceptable level." The compost was subsequently used for landscaping purposes.

Interesting that the mesophilic temperatures were more effective in removing the contaminants, presumably because there is a higher diversity of microorganisms with more "tools" at their disposal. This phenomenon was replicated in a study involving polycyclic aromatic hydrocarbons (PAHs). These are organic pollutants that are widely distributed in the environment, are frequently detected in soils, and are toxic, even carcinogenic. The study proved that both mesophilic and thermophilic conditions were effective in degrading persistent PAHs.

Approximately 30 million pounds of antibiotics are used annually in the US alone for agricultural purposes, about 70 percent of which is excreted in manure. That's an incredible 21 million pounds of antibiotics being released into the environment every year through animal manures in the United States alone! One study showed a 99 percent removal of the antibiotic oxytetracycline after thirty-five days of composting, while less than a 15 percent reduction was achieved at room temperature. After thirty-five days of thermophilic temperatures, another antibiotic, chlortetracycline, was reduced more than 99 percent; the antibiotics monensin and tylosin were reduced from 54 to 76 percent, whereas the antibacterial drug sulfamethazine did not degrade at all in this time period. Another study indicated that composting is effective in reducing the antibiotic salinomycin in manure.

From 2001 to 2003 roughly 3,300 tons of the antibiotic tetracycline

Newly constructed compost bin in Mongolia.

were produced annually for animals in the US. Oxytetracycline is the most widely used tetracycline compound. As an environmental contaminant, it can affect algae, crustaceans, and soil bacteria; can create antibiotic-resistant bacteria; and can risk contamination of the food chain. Approximately 23 percent of the oxytetracycline fed to calves passes through in the manure. Although this antibiotic was present in manure being composted, it did not appear to affect the composting process. Within the first six days of composting, levels of oxytetracycline were reduced 95 percent. The researchers recommended that farmers should be advised of the persistence of oxytetracycline in untreated manure and should compost manure to reduce oxytetracycline residues. In contrast, such residues in manure were not effectively reduced during anaerobic digestion.

When by-products of poppy production were thermophilically composted for 55 days to remove morphine, the morphine content decreased below detectable levels after 30 days, even when the windrow compost was not turned at all. Both male and female human hormones showed an 84 to 90 percent reduction after 139 days of composting in poultry manure. Although the levels of hormones were reduced during composting, they were not completely eliminated during that time period. Perhaps a longer curing phase was needed?

What about residual drugs in animal carcasses? Phenylbutazone (an anti-inflammatory drug) was undetectable after composting. Ivermectin (a deworming agent) had undetectable levels by the end of the composting process. Studies are just beginning to reveal the impact of composting on drugs and drug residues. While more research is needed, recent and ongoing studies are supporting the use of composting.

Much of the pharmaceuticals we excrete leaves our bodies in our urine. A study in Germany recommended not to use urine of people under medication for fertilization of food crops. No doubt they mean direct fertilization. Composting urine beforehand would benefit from the same remediation that is achieved when composting contaminated manure, sludge, and soils.

Then there are the chemotherapy drugs — drugs that directly attack DNA and pass through cancer patients as active chemicals in urine, feces,

Compost toilets under construction in India.

vomit, saliva, and sweat. One of the most powerful and dangerous chemo drugs is cyclophosphamide. Accidental contamination by this drug can cause cancer, birth defects, miscarriages, leukemia, and permanent infertility. Patients can even develop cancers that don't appear for several years. For example, cyclophosphamide, although used to treat breast cancer, can cause bladder cancer. The American Cancer Society warns that toilets used by cancer patients can be hazardous, as can even the lips of a chemo patient — they recommend no kissing. Chemotherapy drugs can exit cancer patients as active and dangerous chemicals. Septic systems and wastewater treatments plants can't remove 98 percent of them, so they end up intact in lakes, rivers, and ponds and eventually into our drinking water supplies. One anti-cancer drug, salinomycin, was composted in manure, and the composting technique was effective in reducing the levels.

Let's hope more research will be conducted in which toxic chemo drugs are subjected to true composting over an extended time period. We can also hope that the medical industry develops treatments for cancer that aren't so threatening and damaging.

Heavy Metals

If the author had a nickel for every time someone asked him "what about heavy metals," he would be living on a fancy private island in the Caribbean. It seems that some composts are made from materials from industrial sites, resulting in some degree of heavy metal contamination in these particular composts. Somehow, this gave the impression to certain people that therefore all compost is contaminated with heavy metals. They then carried this concept one step further to deduce that a compost toilet must also be a source of heavy metals.

So, what about heavy metals in compost toilets? The answer: What heavy metals? Where are heavy metals coming from? If you're excreting heavy metals, then you have a serious problem. Otherwise, how would heavy metals get into your compost toilet?

Heavy metals can be a problem for the commercial composting industry, depending on the source of the feedstocks being composted. Some feedstocks are contaminated with heavy metals such as lead, copper, and cadmium, especially "municipal solid waste" feedstocks.

Soils can become contaminated from the long-term use of public sewage for irrigation because industrial wastes are dumped down sewer drains. Extensive use of chemical fertilizers and pesticides, and careless storage of industrial and mining wastes can also contaminate soils. Your toilet material, on the other hand, is "clean." Unless you're somehow consuming heavy metals.

Studies have found that compost provides a promising strategy to immobilize heavy metals in soils by changing the soil properties. Reactions between heavy metals and organic matter in compost can turn the toxic state of heavy metals into a nontoxic state. Compost applied to contaminated agricultural soils can also reduce the bioavailability of heavy metals, thus reducing harm to plants, animals, and microorganisms. Compost can also reduce heavy metal content in water by 85 to 89 percent through chemical absorption.

Stable, mature, well-cured compost seems to have the greatest effect on binding heavy metals, compared to immature composts. Different metals also react differently to compost; while some may be bound in the soils and kept out of the plant material, thanks to the compost, other metals in the same soil may become more available to the same plants. Plants that tend to accumulate heavy metals can strategically be used to extract the metals from the soils, after which the plants can be safely disposed of, effectively reducing the heavy metal contamination in the soils.

Your body is not a source of heavy metal pollution, so don't worry about your compost toilet. Worry about compost made from heavily contaminated source materials such as municipal solid waste, especially if derived from industrial areas. The solution is to separate discarded materials so trash or effluents containing heavy metals can be quarantined and disposed of or recycled properly.

PART TWO
Chapter Eleven

AFRICAN PRISON
Case Study Review

The Nakapiripirit low-security prison in Karamoja, northern Uganda, Africa, lacked a working sanitation system. The standard toilet in this area is a pit latrine, a hole in the ground where human excrements are directly deposited by individual users. The latrine may be enclosed inside a small building for privacy and for protection from the elements.

Due to the soil structure in Nakapiripirit, the pit latrines were prone to malfunction. The side walls would collapse, rendering the pit latrine useless. The pits would also flood, or fill up with excrement. Emptying them was difficult and quite unpleasant.

Compost toilet systems were installed in 2017 by GiveLove.org, coordinated by Project Manager Alisa Keesey, in cooperation with Samuel Autran Dourado, and with WeltHunger Hilfe.de, one of the largest private aid organizations in Germany.

Prison cells at Nakapiripirit Prison in northern Uganda, Africa, consist of round metal buildings with conical metal roofs. Each can sleep as many as 30 prisoners, all on the floor. Compost toilets replaced bucket toilets in these cells, yielding a far more pleasant living experience.

Men croud into small metal prison buildings, where a compost toilet is a luxury.
Photos both pages by Samuel Autran Dourado.

The prison housed fifty-one men, five women, and two babies. Up to thirty prisoners slept on the floor of a prison cell, consisting of a circular metal building. One compost toilet was placed inside each cell. Prisoners slept on the floor immediately beside it. Prior to the compost toilet, the prisoners used a bucket toilet, an open container of excrement without cover material, which smelled very bad and attracted flies.

The switch from a bucket toilet to a compost toilet was "much better," according to the Officer in Charge, Patrick Emojong. "Before I got the compost toilets, it used to be very difficult," he stated. "I had succeeded to put up one pit latrine, but it was a small one and it would get full over time. Emptying it was a problem."

He added, "Getting cover material becomes a bit of a challenge. A challenge yes, but we meet it, because when we compare it to the system we had before, this is much better. The technology has gone a long way to helping us here. The police need this. The army needs this. The schools need this. This place was supposed to be closed because of poor sanitation!

Officer in Charge Patrick Emojong, inspects a compost toilet in a prison cell. "I am here because of this technology," he asserted. "By now, I am assuring you, this place would have been closed." Photo by Samuel Autran Dourado.

I am here because of this technology! By now, I am assuring you, this place would have been closed."

The Officer in Charge further elaborated, "This can go a long way in alleviating the sanitary problems in this community, and in places like hospitals, and households."

What do they need to keep a successful compost toilet system functioning at Nakapiripirit Prison? For one, the prison would benefit from hand-held sprayers for spraying water to clean toilet receptacles. The ones they were using were wearing out. They would also benefit from a chipper/shredder machine to enable them to generate their own cover material from surrounding vegetation, rather than having to search for factories to get sawdust, rice husks, and coco coir. The local farmers are

All of the prison's toilet material is composted in wire bins and covered in local grasses. The round metallic building in the background is a prison cell inhabited by 20 cellmates. The prison was facing closure due to poor sanitation — the pit latrines were collapsing due to the loose soil structure. Bucket toilets were employed, but they were disgusting. Compost toilets saved the prison and greatly improved the quality of life.

needlessly burning the ground cover. Gathered up and ground in a machine, that ground cover could instead create one of the essential components of a compost toilet system: the cover material.

People in developing nations with little or no sanitation options can easily create compost toilet systems for prisons, schools, villages, orphanages, hospitals, and households, if they have a little help. Even shovels or other hand tools come in handy. Wire fencing for bins, chipper/shredders or other grinders for cover material, spray bottles for cleaning toilet receptacles, long-handled cleaning brushes, small hand brooms for brushing up stray sawdust, and other accoutrements of the compost toilet phenomenon are all immensely helpful to people of meager means.

According to Oxfam (2020), "Hundreds of millions of people are living in extreme poverty while huge rewards go to those at the very top. There are more billionaires than ever before, and their fortunes have grown to record levels. Meanwhile, the world's poorest got even poorer. Many governments . . . are massively under taxing corporations and wealthy individuals yet under funding vital public services like healthcare and education. These policies hit the poor hardest. The human costs are devastating. . . ."

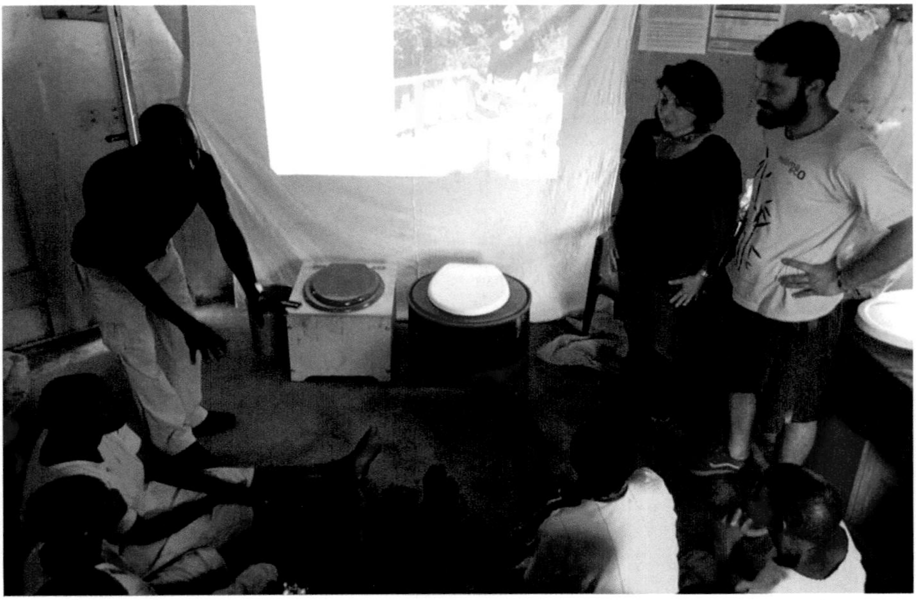

Prisoners are instructed how to use and manage the compost toilet system.

Chapter Twelve

MONGOLIAN GER COMMUNITIES

CASE STUDY REVIEW

Mongolia is a landlocked Asian nation wedged between China and Russia. The capital, Ulaanbaatar, is considered one of the coldest capital cities in the world. Making compost in Mongolia seemed like the ultimate challenge. Here we have a very cold, dry climate; little vegetation available; almost no food scraps (the dogs eat it); very poor people; and a nomadic culture with almost no history of gardening or composting.

Rainfall only amounts to about 10 inches per year (25 cm). The population of the entire nation is only about three million, scattered over 1.5 million square kilometers (about 600,000 square miles), with a total population considerably less than the city of Los Angeles, California, USA.

The country is covered primarily in steppe, vast expanses of grassland without trees or even shrubs. The people have a history of nomadic herding, living off their sheep, goats, horses, camels, yaks, and cattle, moving from place to place as the grazing allowed. Their round houses, known as *gers*, but often referred to as yurts in other countries, are made from the very hair of the animals in their herds, felted into mats and draped over wood frames. The houses could be collapsed at will and loaded onto a horse-drawn wagon, to be moved from place to place as needed.

Because of their nomadic culture, land ownership was a foreign concept for many generations. Gardening was not a common practice. Composting was largely unheard of. Meat from the herds, along with fermented horse milk, were staple foods. The one-room round gers nurtured an open, welcoming culture. Mongolians never failed to be extremely accommodating to the author, offering all of their foodstuffs at every ger visit. Even vodka was served at breakfast.

In the early twenty-first century, extreme weather and heavy snowfalls were starving the herds. The animals could not access the grasses buried underneath the snow. Nomads flocked to the cities by the hundreds of thousands to seek out an alternative lifestyle. They set up their gers around

Other than grasses, the Mongolian steppes lack vegetation.

the main population centers. Each family was allotted a small piece of land, which they enclosed with wooden fencing. Almost all ger household plots had at least one dog, whose only food source was the scraps that came from the household kitchen. This left little to be composted.

When the author landed at the Ulaanbaatar airport in 2006 to introduce Mongolia to compost toilets, it was –15°F (–26°C). Nevertheless, a two-phase feasibility study was successfully developed from late February to mid-May that year, funded by the Asian Development Bank. The purpose of the study was to determine whether compost toilet systems could provide a successful alternative to the pit latrines currently in widespread use in the ger communities.

A PowerPoint presentation was shown to the Ministry of Construction and Urban Development in the capital city Ulaanbaatar (UB), with government representatives, members of NGOs, and other interested persons attending. The presentation was then shown to larger groups of community members, *bag* (neighborhood) leaders, and ger inhabitants, both in UB and another city, Erdenet. The presentation received coverage on national and local TV and radio. Two compost toilets were subsequently constructed by ger inhabitants, the process recorded and broadcast by a TV crew.

The compost concept proved to be new to the general population, as did the entire compost toilet idea. However, the reaction to the presentations was surprisingly positive. In fact, numerous ger inhabitants expressed a desire to immediately begin using the toilets.

This is not surprising considering the fact that virtually all of the ger inhabitants were using pit latrines— small outdoor shelters with minimal privacy in many cases, and a space in the flooring boards over which one squats when relieving oneself into an open hole in the ground. The floorboards are often splattered with urine and perhaps fecal material, frozen during the winter months, creating a slipping hazard.

If the ger population were to be connected to the municipal water sources and their wastewater drained into the sewage system, the existing wastewater treatment plant would have quickly become overloaded. Yet environmental issues associated with the increasing number of pit latrines were becoming a cause for concern. In Ulaanbaatar, the drinking water

Continued on page 172

This is a pit latrine at a public restaurant in Mongolia. Urine freezes on the floor of these pit latrines, creating a very slippery and unpleasant surface on which to squat.

Although these pit latrines look primitive, it is not unusual for finely dressed ladies in high heels to have to use them in Mongolia, even when frozen, as shown above. One can easily imagine how indoor compost toilets could greatly improve their lives.

supply originates from groundwater sources, some of which are located near or underneath the city and may be as shallow as 2 meters (6.5 feet). Precipitation doesn't provide enough water for groundwater recharging. With nearly six hundred thousand people using pit latrines in the area immediately surrounding Ulaanbaatar, the situation would soon become a water pollution time bomb.

The approximately 120,000 pit latrines, most dug to a depth of about 2 meters, were constantly filling with human fecal material. This pollution was slowly working its way into the underground water table. Once the groundwater becomes polluted, there is little that can be done to reverse the pollution other than to remove the many sources of pollution and then allow nature to run its course.

In short, the potential for fecal contamination of the drinking water supplies in the Ulaanbaatar area was alarming. Compost toilets can remove that source of contamination depending on to what extent the toilet systems are employed.

The prospect of an indoor toilet without odors that would not pollute the environment, but which would produce soil fertility, was attractive to many people. Immediately after one of the first public compost toilet presentations, which included photos of the author's gardens, an elderly lady approached him with a note, written in Mongolian, asking for help in improving the soil fertility on her plot, writing, "I want to make my plot just like yours." At another public venue, a group of about ten ger ladies who had attended the presentation waited outdoors in the cold for quite some time just to tell the author how grateful they were to have this information and how they would "work so hard" to be successful at using a compost toilet. One elderly man at a public presentation, during the question-and-answer period at the end, stood up and suggested they begin using the toilet systems immediately and that they have a competition between neighborhoods to see who can make the best compost.

Little technology is required for the operation and long-term use of compost toilets. The costs for constructing both the toilets and the compost bins are low. The primary costs lie in the production and transportation of organic cover materials. However, when the cover materials consist of

There are billions of people on planet Earth who do not have the luxury of running water in their homes. This is troublesome in Mongolia, where water pipes must be buried 3 meters belowground to be protected from freezing. Water is transported by handcart from tank trucks to houses in the ger communities. No flush toilets are possible.

organic materials from such activities as the milling of lumber or the production of crops or food by-products, the use of these materials may be available free and may not require a financial expenditure other than to transport the materials to the ger communities.

In some cases, a business may actually pay to have discarded organic material removed from their business site, thereby offsetting the transportation costs. For example, the Erdenet carpet factory paid about $30 US per day to have 9 cubic meters of scrap wool transported to a landfill. They were willing to pay the same amount of money to have the material hauled away for composting purposes. Wool waste accumulates when the incoming natural wool is precleaned prior to entering factory production. It contains quantities of both manure and dirt, which makes it attractive as a composting feedstock.

Compost toilet systems can either be centralized or decentralized. Decentralized, or "on-plot" systems, when the compost is confined to the individual household plot, can be created and maintained by individual households without the need for government control, oversight, or expense. Centralized compost systems, on the other hand, involve a scenario where village household toilet containers, when full, would be routinely collected and removed to a centralized composting facility by an outside service. Such a service-oriented system may require some degree of training and oversight.

In an effort to gauge the nature and quantity of organic cover materials that may be available for use in Mongolia, sources of such materials were sought out. The local rug factory was visited, as were the paper factory, abandoned lumber mills, local sawmills, and the government agricultural department. Significant quantities of sawdust were located.

Sawdust is produced in both Ulaanbaatar and Erdenet as a by-product of the local lumber industry. All of the ger area household plots are encircled by wooden fencing that was sawn from trees at a mill. In addition, many of the household dwellings are constructed of wood, which is harvested locally. Although the steppes are barren of trees, as are the dry south slopes of the mountains, the *north* slopes of the mountains are covered in trees. The sawdust from this lumber can be used in compost toilets if it is

Continued on page 177

These compost toilets constructed in Mongolia in 2006 cost approximately $12 US each, including labor and one receptacle. Below are community residents selecting their own personal toilet for use in their ger household.

Proud owner of a new compost toilet, this retired Mongolian agronomist is looking forward to abandoning his pit latrine and making compost instead. The compost toilet below was built by a municipal official for use in his private home in Mongolia.

made available free of charge or at low cost to the households that elect to use such toilets.

The sawmills in UB were allowing the sawdust to be taken for free, while large sawdust piles in Erdenet appeared to have been abandoned. A logistical challenge involves transporting the sawdust to the ger areas. Most ger residents do not own cars or trucks. In addition, such organic materials as waste wool, grain chaff, and fermentation by-products, could potentially be utilized as cover materials in compost bins.

Where there is no direct connection to piped water, water usage in the ger areas is minimal because the water is moved by hand to the household. A family of two, for example, may only utilize 240 liters (63 gallons) of water per week. Some of that gray water could be used in a compost system for washing toilet receptacles, for example, and for wetting the compost when necessary.

During the summer growing season, if gardens are established, gray water can also be used to provide water for the gardens, shrubs, and trees. Excess gray water could be disposed of down a soak hole during the winter months, similar to the hole utilized in the pit latrines. The gray water should not be disposed of in pit latrine holes because this will increase the likelihood of groundwater pollution by the fecal material leaching from the hole. When a compost toilet is used, a pit latrine is not necessary. A relatively small gray water soak hole can be dug separate from the pit latrine hole. The dirt taken from the gray water hole can be used to cover the pit latrine contents, thereby minimizing the leaching of fecal material while masking the unpleasant odors.

Families who wanted to establish compost toilet systems on their ger plots were required to participate in a training program, which included watching a fifty-minute instructional DVD and reading and reviewing an instruction manual. Volunteers who successfully completed the training were then provided with a compost toilet and were instructed on the construction of a compost bin and acquisition of cover materials.

Due to time constraints and the physical distance between the two pilot project cities (approximately seven hours by car, one way) it was decided to limit the pilot project groups to six families per city, or twelve families

Advancing from a pit latrine to a compost toilet system constitutes a revolution in the lives of many people. Billions of people have never had an indoor toilet, nor have their ancestors, since the dawn of humanity. Where water toilets are impossible, compost toilets, theretofore unheard of, can provide an alternative.

total. The two cities selected were Ulaanbaatar, population about one million at that time, with about 60 percent living in ger communities, and Erdenet, Mongolia's third largest city with a population of about eighty-seven thousand.

Twelve toilets were constructed in one afternoon and provided to the families free of charge. The cost was about $12.00 US per toilet, including labor and overhead. One compost bin was constructed as a demonstration project in UB; then the UB participants each built their own. In Erdenet, the group was shown the photos of the UB compost bin workshop, which provided enough information for them to build their own household compost bins. The demonstration bin in UB cost $13.00 using new materials. Most participants used salvaged or repurposed materials to build their bins.

Government representatives expressed strong support for the compost toilet project, including the governor, deputy governor, and cabinet in Erdenet. A meeting was held at the Erdenet government building with the director of the Department of Agriculture, including the area agronomist and a representative from the Department of Health. This group possessed a scientific understanding of composting and were aware of the proven validity of the process with respect to pathogen reduction. They also concurred that human manure must be recycled for agricultural purposes sooner or later, and that the soil amendments traditionally available in Mongolia are diminishing due to the reduction in the number of herd animals and animal manure.

Dried animal manure is the primary source of fuel for the ger fires, used both for heating and cooking. The resulting loss of soil fertility could be offset by recycling human manure. Human manure compost could become valuable enough to buy and sell, thereby creating an economic market theretofore unavailable. Government representatives recommended the establishment of a permanent laboratory for the purpose of analyzing finished compost from compost toilet systems to monitor for potential human pathogens and to assess the agricultural quality of the compost.

The problems most likely to occur with inexperienced composters include a) the compost being too dry and therefore not developing internal heat; b) the compost being frozen for a long period and the compost bin

Proud owners of newly built compost bins in Mongolia: the crew at top used wood boards; the ladies at bottom used repurposed steel and wood posts.

therefore becoming too full (not shrinking), or the compost pile too tall (not flattened); c) insufficient cover material being used, which will be indicated by odor or leachate; d) dogs or other animals getting into the compost (indicating a need for a better compost bin construction); e) flies on the compost pile (indicating a need for more cover material); f) irregular use of the toilet system, not allowing for sufficient mass to accumulate in the compost pile; and g) improper cover materials such as wood chips, bark or other larger wood chunks, or dirt, wood ashes, or coal ashes.

The household compost toilet systems require approximately a year of collection time of the compostable materials, followed by another year of curing during which the compost is left undisturbed. As such, there is a minimum time period of approximately two years before the first finished compost is available for analysis.

The cost to set up one compost toilet and compost bin, using new materials, was about $25.00 US per system (2006), not including cover materials, except one bale of hay at $1.50 US. Sawdust was free.

These were actual costs determined by buying the materials in Ulaanbaatar at the construction markets. The toilet receptacles and cleaning brushes were purchased at the "Black Market." Costs can be significantly reduced by using salvaged materials. Although a demonstration compost bin was built in the UB area out of new wood, the pilot project participants built their own bins out of the materials they had on hand, for little or no cost. The toilet cabinets can also be constructed from recycled materials, although ideally, they would be mass-produced to make sure the details are correct. For example, toilet receptacles must closely fit the toilet cabinet hole; the hole for the receptacle must be set back only about 3.8 cm (1.5 inches) from the front of the toilet cabinet; the height of the toilet box must be correct; and so on.

An individual compost toilet system should utilize more than one toilet receptacle. At least four are recommended for convenient use of the toilet. That way, the toilet can continue to be used even after one receptacle or even three are filled because the filled ones can be set aside, with lids, and emptied into the compost bin at the convenience of the user without interrupting the use of the toilet.

This entire project was abandoned, due to the exhaustion of funding, in 2006. The author, at his own expense, returned to Mongolia in the summer of 2007 to conduct follow-up research. However, none of the pilot project ger residents could be found. As is their custom, they had all moved out to their summer gers away from the city and could not be contacted.

The Mongolian government could promote composting as a sanitation alternative by having cover material delivered and dumped in piles at designated "compost resource centers" in ger communities for free access by composters. They could offset this cost by taxing pit latrines. This would discourage pit latrine use and encourage compost toilet use. On the other hand, why tax the poorest of the poorest? How about taxing the people who are hoarding wealth, and then use some of *that* money to help those who need it? What's required is political will, which is something that seems hard to come by when the well-being of poor people is concerned.

In 2006, the average Mongolian was earning $2 US/day while the richest 1 percent of the world's population owned 40 percent of all global wealth.

Large amounts of sawdust from lumber production in Mongolia remained unused, even abandoned. This is perfect cover material for compost toilets, although most people in ger communities don't have cars or trucks for hauling it.

Chapter Thirteen

HAITIAN SCHOOLS AND ORPHANAGES

GIVELOVE.ORG PROJECTS: CASE STUDY REVIEWS

AMURT SCHOOL

The Amurt School in post-earthquake Port-au-Prince, Haiti, was a "green" school consisting of eleven pavilions, twenty-two classrooms, eight compost toilets, eight rainwater catchment systems, a reservoir, two composting sites, a tree nursery, a permaculture demonstration site, and organic gardens. Two agronomists and three technicians provided classes and demonstrations of urban permaculture, focusing on the 820 children who attended the preschool and after-school programs, plus the parents, and the women's and youth groups.

Students utilized two toilet buildings, each with four toilet stalls. The private toilets had chutes through the floor and receptacles underneath, each with a capacity of approximately 60 liters. They were easily replaced

Continued on page 188

This is one of the Amurt's four-stall toilet buildings. Note the proximity of the compost bins at right. Transport costs are eliminated when the bins are near the toilets.

The older compost bins at Amurt were still reading 131°F (55°C) six months after filling. This is one reason the compost should completely cure before using. Immature compost will kill plants. Allow the compost to cure for approximately one year.

The Amurt toilets used chutes through the floor which directed toilet material into 60-liter containers underneath. A handwashing station is attached to the building.

The school's parents and supporting community at Amurt are shown the process of "center feeding" the toilet material into the compost pile.

The piles are regularly monitored by the compost crew.

with empty receptacles when full. Sugarcane bagasse and amyris wood sawdust were utilized as cover materials. The system produced approximately eight 60-liter containers of toilet material every three days. This volume filled a compost bin measuring 1.5 meters wide, 2.0 meters long, and 1.0 meter high (5' wide×6.5' long×3' high) every three months.

Food scraps were added to the compost bin twice a day. Temperatures in the nearly full active bin were hovering at approximately 140°F (60°C) when monitored by the author. A bin that had already been filled and had been aging for six months was still reading 131°F (55°C).

The compost bins were located immediately adjacent to the toilet stalls, allowing the contents to be conveniently composted without the need to transport the toilet material. The system had no unpleasant odor; you could have a picnic beside the compost bins. Flies were not evident. Hand-

Above: Cured compost is used on the Amurt school grounds to grow plants. **Opposite page:** The Amurt garden campus prior to its destruction.

washing stations were conveniently located next to the toilet buildings.

The self-managed compost toilet system has a dedicated crew of three persons: one woman who cleaned the toilet buildings and receptacles and two men who managed the composting, emptied the receptacles and kept a supply of cover material. They were all paid by the school.

Posters explaining the proper use of the toilets were displayed in the toilet stalls. Community training seminars were provided by GiveLove.org.

Speaking before the United Nations General Assembly in New York, the Haitian minister of education received applause for this incredible school. The minister proclaimed that in less than two decades the educational model of Amurt would be standard throughout Haiti. However, in the early morning of September 5th, 2014, *"parents and children watched in horror as bulldozers and other heavy machinery destroyed this model school. The landowners – one of the richest and most powerful Haitian families – had won a court case against other former owners. Months of negotiations with the Amurt school sponsors to purchase the plot of land where the small school stood had apparently been merely a diversion tactic by this family. Ultimately, Haitian Prime Minister Laurent Lamothe signed a decree to remove by force this school ..."* (from the kindernothilfe.org website).

When the author returned to the site to conduct follow-up documentation, there was no trace of the school. Instead, a wall surrounded an empty property. What was once an excellent demonstration site for compost sanitation had suddenly, and sadly, vanished.

Orfelinat Enfant De L'Espoir or "Children of Hope" Orphanage

The Orfelinat Enfant de L'Espoir or "Children of Hope" Orphanage in Leogane, Haiti, has a compost sanitation system very similar to the Amurt School. The toilet stalls are located near the compost bins; inside each stall is a poster describing how to use the toilet. There is a rainwater catchment system attached to the toilet building, which provides water for washing toilet receptacles and for handwashing. Both 20-liter and 60-liter receptacles are used — 20-liter toilets for the smaller children and 60-liter toilets for the bigger children and adults. The toilet building is adorned with a painting of the "human nutrient cycle." The 20-liter receptacle lifts out the top of the toilet cabinet, while the 60-liter receptacle slides out the front. The entire process is odor-free.

Children of Hope Orphanage in Haiti had pit latrines converted into compost toilets. The toilet at left is for the smaller children. The toilet building, compost bins, and handwashing station are shown above.

The Sopudep school compost toilet building is adorned with a painting of the "human nutrient cycle." The compost bins are nearby, allowing the children an opportunity to see the compost operation in progress. The process is odor-free.

The Sopudep School in Petion-ville, Haiti

Harry Ha of Partners with Haiti and the Toronto Haiti Action Committee traveled to Haiti in February and March 2011 to lead the installation of four compost toilets at the Sopudep School. GiveLove.org aided their endeavors. The sanitation system utilized 20-liter receptacles and was self-managed by school personnel, serving five hundred children when school is in session. The stall has a painting of the human nutrient cycle on the outer wall for educational purposes. The compost bins are immediately adjacent to the toilet building and the bagasse supply is nearby. Bagasse is also stored next to the toilets in wall-mounted dispensers. Rainwater is collected and provided for handwashing.

Potential Problems with Compost Sanitation Systems

Inadequate information, training, understanding, and knowledge, compounded by fear of fecal material, can make compost sanitation systems flounder, or fail. People who are not serious and conscientious with alternative sanitation should not be managing compost toilet systems. There should be a dedicated, trained compost crew that understands and cares about what it's doing.

Correct and adequate cover materials must be used in adequate quantities. If bad odors are smelled, or flies are attracted to the toilet or the compost bins, then more cover material must be used.

A mismanaged compost pile at an American relief agency site in Port-au-Prince used inadequate cover material. Organic material was simply cast onto the pile without digging a depression first. There should be nothing visible but the cover material — no flies, no maggots, no food, no toilet material — and there should be no unpleasant odor. The pile was odorous and was breeding maggots when it could simply have been covered over adequately and these deficiencies avoided. The problem at this site was caused by a frequent turnover of volunteers such that a dedicated compost crew was not available. Continuity of training was not maintained, and the proper maintenance of the compost system fell to the wayside. Although the system had been set up by trained and knowledgeable personnel, the compost toilets were eventually removed from the site by the same personnel, due to user neglect and mismanagement.

Toilet wash water must go into the compost piles. Compost toilets must be kept in a heated area in cold climates; otherwise, the contents of the receptacles can freeze and will not be able to be emptied. Frozen receptacles can also crack and leak. Compost bins are sometimes located too far from the toilet area. This can make the job of emptying toilet receptacles burdensome and unpleasant if the toilet receptacles are managed by hand. There is no reason to locate the compost bins far away. If they are correctly managed and covered, there is no odor. People who fear human excrement assume the compost bins will smell bad, and they will put them so far away from the toilet as to be impractical. This encourages failure. The solution is to use enough cover material in the bins to prevent odors from escaping no matter where the bins are located.

A poorly managed compost pile at a relief agency site in Port-au-Prince, Haiti, has inadequate cover material on top. There should be nothing visible but the cover material — no flies, no maggots, no food, no toilet material — and there should be no unpleasant odor. This pile was odorous and was breeding maggots when it could simply have been covered adequately instead. The problem at this site was caused by a frequent turnover of volunteers such that a dedicated compost crew was not maintained. Continuity of training and the proper maintenance of the compost system fell to the wayside.

Chapter Fourteen

SANTO VILLAGE, HAITI

GiveLove.org Project: Case Study Review

Two US-based charitable organizations collaborated with the Leogane community in Haiti to create a village for housing people left homeless by the 2010 earthquake. Five hundred homes were to be constructed, each on a 150-square-meter lot. Each home was expected to have its own toilet to avoid the maintenance concerns associated with shared sanitation systems. The primary objective was to establish a system with relatively low maintenance, cost, and labor, based on a reliable and proven technology.

For practical considerations, one of which included poor access to water on the site, the options were narrowed down to: a) Compost Toilets; b) Enclosed Long-Term Dry Toilets; c) Pour Flush Toilets; d) Pit Latrines; and e) Urine Diverting Dry Toilets

Of the five options, Enclosed Long-Term Dry Toilets (ELTDT) was selected. The commercial system had been developed in Namibia, Africa, a country that receives about 14 inches of rain annually (356 mm). Leogane receives about 53 inches (1347 mm) annually. In hindsight, this should have been a cause for concern. Reasons cited for choosing this system included advantages in maintenance, operational costs, social acceptability, health risk, reliability of the technology, environmental factors, and economics. The cost for this system was over a half million US dollars.

The Compost Toilet System, priced at about one-third the cost, had its drawbacks, according to the developers, who stated, "The social acceptability of this [compost toilet] system would probably be quite low as there is repeated contact with the humanure. Having to store humanure for collection may also be undesirable."

An additional concern to the developers was health and risk. They stated, "There are large numbers of people coming in direct contact with the humanure, increasing the risk of contamination by pathogens. The toilet seat may have to be secured to ensure children cannot gain access to the humanure. Storage of humanure in the household prior to collection is a potential risk, as is securing the bins when left out for collection. Compost managers will need to ensure compost stations are also kept secure to ensure they are not accessed by unauthorized personnel. All tools and transport associated with the composting stations will have to be cleaned properly after use. Compost station staff will need to be very well trained and know what measures to take if temperatures are not being maintained according to guidelines."

The chosen ELTDT system was described as *"a dry toilet system which collects humanure in a large, perforated bucket in a chamber beneath the toilet. A specially designed urine diversion toilet is used to divert 80% of the urine directly into a soak pit in the ground. This type of urine diversion toilet does not need special instruction for use. The remaining urine and solids are collected in the 90-liter bucket below. The remaining urine percolates through the solids and seeps into the soil. The bucket for a single-family use usually fills up within 6 months. It is removed to the rear part of the chamber below the ventilation chimney to dry out the remaining solids and replaced with an empty bucket. When the second bucket fills, the dry solids are removed and returned to a hole in the soil, unless further composted. A black painted panel and large chimney vent help to remove smells from the toilet. Orientating the ventilation stack towards the sun is important for efficiency. On occasion this system has been installed with a small solar-operated fan for night ventilation."*

A 90-liter container of water will weigh about 200 pounds (90 kg). A 90-liter container of toilet material would be very difficult to move and empty by hand, especially when located underneath a toilet building. Furthermore, before the container contents can be emptied, a hole must be dug in the ground where the toilet contents can be dumped, which makes this system seem almost like a complicated pit latrine system. Unless, of course, the contents are simply dumped on the surface of the ground, which would be the easiest way to dispose of it.

The ELTDT dry toilet was considered "ideal for dry and hot climates." Even though Haiti tends to be much wetter than Namibia, the system was installed anyway and in use for a period of time at the village. However, the villagers soon complained about unpleasant odors emanating from the toilets, which hung in the air throughout the village. Residents had to close

This is a USAID pit latrine at the preliminary Santo Village in Haiti. The toilets are so rank and disgusting that children would rather defecate outside on the ground than go inside and be assaulted by the stench and the flies, or risk falling into the hole. Pit latrines like these can be modified into clean, odorless, compost toilet stalls.

their windows at night, despite the oppressive heat, to try to keep out the odor so they could sleep. The toilets became so unpleasant that villagers refused to use them, some resorting to open defecation. The ELTDT dry toilet system eventually was completely abandoned.

A review of the system yielded the following information:

1. The toilet contents had a soupy consistency that was very wet, odorous, and unpleasant.
2. When full, the 90-liter containers of raw excreta were expected to be lifted out of the sunken toilet chamber, removed to another location, and safely buried according to the discretion of the user.
3. There were unknown expenses for the toilet user related to the removal and burying of the excrement. Costs included labor, transportation, and possibly a disposal fee.
4. Wet fecal slurry was in and on the dry toilet drums and around the chamber floors. The chambers omitted strong odors. All the 90-liter containers observed were covered with notable insect infestations, flies, fly larvae (maggots), and feces.
5. The odors and flies were probably the result of the spillage or

The "Enclosed Long-Term Dry Toilet" system utilized space underneath a toilet stall where a container was located for the collection of sewage. Odors, flies, and the difficulty of emptying the container led to the system's abandonment. Now the toilet stalls house small-footprint compost toilets, with the remainder of the building used for storage.

Compost toilets require little space. The Santo Village compost toilet system collected toilet material from over 250 households, each with their own private toilet building. The system had no odor at all and produced literally tons of high-quality compost.

overflow of feces and urine in the chamber due to a drop shoot that was too long or not designed properly for the toilet contents to fall entirely into the drum.

6. The fly infestation could pose a health risk since flying insects can spread disease. The uncleanliness of the chambers and the hundreds of flies presented a psychological barrier for the users. The toilet chamber was very unpleasant and therefore less likely to be cleaned or maintained.
7. Users were pouring bleach and other cleaning products into the toilets to control odors. They complained that the toilets "smelled very bad at night in the house," despite their location outside.
8. A user reported that her family of five had already filled one toilet drum in only two months of use.
9. A user reported that they had exchanged the 90-liter drum for a smaller container so they could dump it more easily.
10. A user said that they were going to stop using the toilet because they would rather go outside than deal with the heavy containers.
11. Several users reported that they were concerned about the flies in the chamber and did not want to touch the containers at all because they were very dirty.
12. Users expressed fear about disease and cholera, fear of moving the large toilet container, or going into the sunken chamber to clean it.

Other concerns listed included:

1) The excreta in the toilets didn't dry sufficiently in Haiti.
2) The receptacles could overflow when in use.
3) The toilet material was not pathogen-free.
4) A proper maintenance schedule was not developed.
5) People were opting out of using the toilet because the odors and insects were not controlled.
6) Users were trying to dispose of the toilet contents themselves, creating a health hazard, a source of fecal contamination in the environment, and community conflict when they dumped untreated sewage.

7) Chemical additives such as motor oil and bleach used to control odors were causing environmental pollution, as well as rendering the toilet material difficult to compost and unsafe for agriculture.
8) Land in or near the village was needed to bury toilet material from five hundred homes four to six times a year.
9) There was a high risk of failure when the toilet maintenance and disposal was left to the user.
10) Cleaning the chamber exposed household members to direct contact with untreated feces.
11) The feces/urine slurry in the toilet drums could not be applied to soil or used for agricultural purposes due to pathogens.

The expensive ELTDT system had failed dramatically. Once again, interest turned to compost toilets. When the compost toilet alternative was revisited by the village community, they voted to switch. The compost toilet project was soon implemented and scaled up to nearly three hundred homes over the course of a week. The toilet buildings became housing for the compost toilets, with plenty of room left over for general storage. The new compost toilet system had been in use at the Santo Village for nearly three years when the data for this account was collected by the author.

Two composting sites were built in the village, each enclosed by chain link fence and razor wire to keep out vandals. Inside these compounds were sheds with padlocked doors for storing equipment and materials. Water wells pumped water into large aboveground tanks for cleaning purposes. Gravel "soak pits" were installed for disposing of the final rinse water after cleaning toilet receptacles. Compost toilets were set up inside the compost yards for workers. The compost yards allowed access to large trucks for the delivery of bagasse, which was delivered every six to nine months and dumped in large piles inside the fence. The bagasse was free, but the hauling was expensive. The cost of hauling bagasse in rental trucks constituted about one-third of the operating costs of the entire compost toilet program.

Sixteen compost bins were constructed in each compost yard, for a total of thirty-two bins, all constructed of wooden pallets laid on edge. Each bin had a holding capacity of 8 to 10 cubic meters, and each bin took approximately one month to fill with toilet material from 1,000 to 1,400 users

The Santo Village was a group of small houses (above) designed and built for Haitians displaced and left homeless by the 2010 earthquake. Ground sugarcane stalks ("bagasse") make an excellent cover material in compost toilets and bins. Below, a truckload was dumped at the Santo Village for use in their community compost toilet operation. Although the bagasse was free for the hauling, the cost of hauling it was a primary expense for the composting operation. **Opposite page**: finished compost is admired (top), while working compost bins are probed for temperature readings (bottom).

(1,500 to 3,000 according to the compost manager). Every bin was labeled by number, with additional information, such as the date they had been started, the date they had been completely filled and closed off, and the date due for harvest.

The village of two hundred households (270 toilets according to the compost manager) produced approximately 2,800 pounds (1,270 kg) of finished compost per month on average, or 33,750 lbs./year (15,309 kg), or approximately 17 tons/year (15.3 metric tonnes). This amounted to 62.5 cubic meters or 1,125 thirty-pound bags of finished, cured compost, enough to supplement their garden plots and orchards with soil fertility, or to sell for money.

It is important to note that, after the compost toilet system had been implemented, there was no sewage produced in this village whatsoever. There was no odor at the compost yards, nor at the toilets or toilet buildings. Flies were rare. This is in direct opposition to the ELTDT system, as well as to "Pit Latrines," "Pour Flush Toilets," and "Urine Diverting Dry Toilets," all of which are disposal systems, not recycling systems like compost toilets.

Let us dwell on this issue for a moment. This seems to be one of the hardest concepts for people to grasp about compost toilets. They are not collecting "waste" for disposal. Compost toilets are collecting organic material for recycling. Once recycled, that organic material has been converted into a valuable resource: compost. The compost is then used to grow plants. Whether they are food plants, fiber plants, decorative plants, or feed plants for livestock doesn't matter. Compost has value; that's why we make it. We can eliminate sewage and the stench, pollution, disease, waste, and other negative consequences of sewage production, and instead replace it with odorless, hygienically safe, beneficial compost, if we understand the practical processes and know how to do it.

To start the process at the Santo Village, first, a small compost team was trained. Each household was then provided with 20-liter (5-gallon), waterproof, plastic toilet receptacles, with lids, for the collection of the toilet material. Receptacles this size can easily be handled by one individual. The receptacles were housed in wooden cabinets from which they could

Continued on page 207

The compost area quickly grew up in banana trees, creating a banana garden. Every bin was labeled with important information, such as when to harvest the compost.

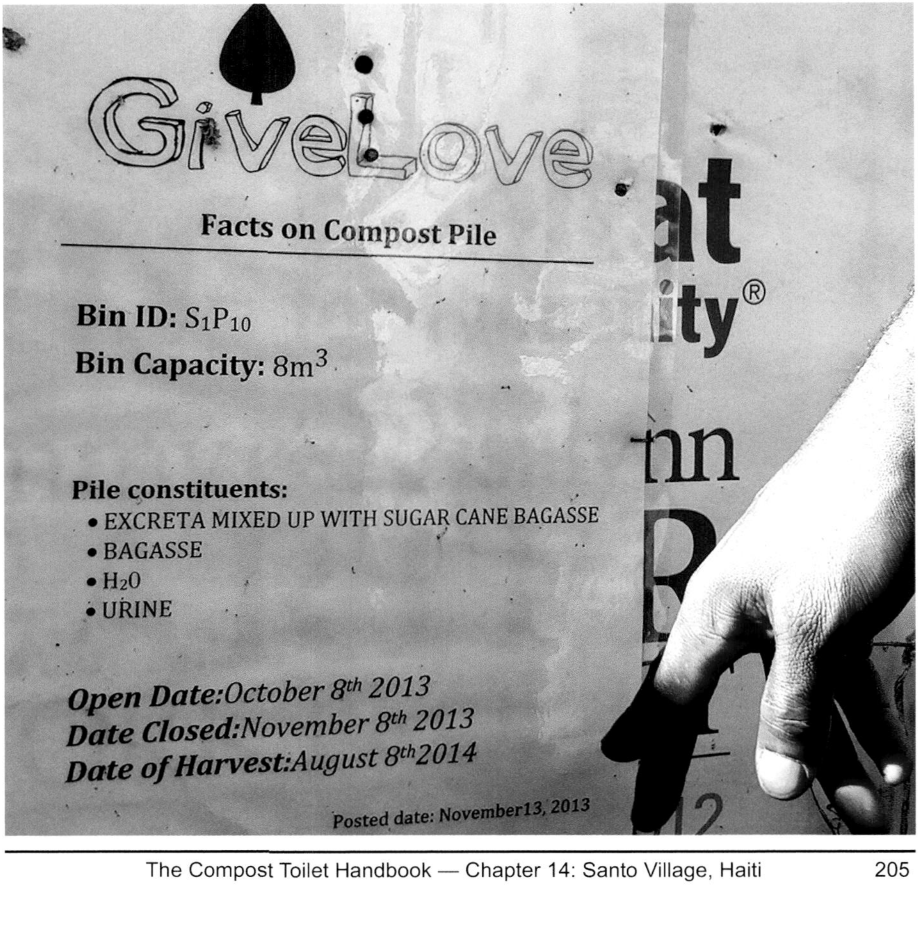

GiveLove
Facts on Compost Pile
Bin ID: S1P10
Bin Capacity: 8m3
Pile constituents:
• EXCRETA MIXED UP WITH SUGAR CANE BAGASSE
• BAGASSE
• H20
• URINE
Open Date:October 8th 2013
Date Closed:November 8th 2013
Date of Harvest:August 8th2014
Posted date: November13, 2013

Finished compost is used in a test plot to grow pepper plants (above).
The compost yard is fenced but allows for truck access (below).

be easily removed. A toilet was located inside every repurposed toilet stall located behind each home, although the compost toilets *could* have been moved indoors since they had no odor. This is one of the main advantages of compost toilets: they can be located anywhere that is convenient, including at the bedside of the elderly. It is also one of the hardest concepts about compost toilets to grasp by people unfamiliar with their use, especially people who grew up with pit latrines or other smelly, fly-infested, maggot-breeding toilets. To them, a toilet is something that smells bad and needs to be located away from the house. Not so with compost toilets.

The Santo community received training on how the compost toilet system works and how to use it. Instructions were posted on walls adjacent to toilets. After a toilet was used, the user simply covered the contents of the toilet with sugarcane bagasse. When anyone looked inside a toilet, all they should have seen was the bagasse (and maybe some stray toilet paper). No urine was diverted, and all paper products were included in the toilet. By managing the toilet in this manner, no odors escaped, and no flies were attracted to the toilet contents.

When a toilet receptacle became full, the user simply lifted the 20-liter container out of the toilet and set it inside the toilet room, with a lid on it. Thereby, all toilet material was collected, including fecal material and urine, and nothing was leaked, drained, or dumped into or onto the soil. There was also no human contact with excrement, contrary to the fears of the original developers who had an extremely poor understanding of compost toilet systems and how they work. And unlike the ELTDT system, nothing from the compost toilets was polluting the environment or creating a health hazard.

Twice a week, as needed, the toilet user would carry his or her toilet receptacle, with lid, to one of the nearby compost yards, early in the morning when it was still cool outside. This was about a five-minute walk. Here, inside the fenced compound, they would simply set their toilet receptacles on the ground for the compost workers to handle. The villagers would then take clean, empty receptacles as supplied by the compost workers, put clean bagasse in them if needed, and carry them back to their homes. At the time of this writing, there were two of the author's videos on YouTube showing the "Village Compost Toilet System in Santo Village," if you want to see the actual site.

The compost workers would then empty all the toilet receptacles at once into a designated compost bin. Approximately 150 to 200 receptacles were collected twice a week, and a single bin was filled in approximately a month's time. The author visited the site on a day when the toilet receptacles had been emptied that morning. There was no odor whatsoever.

Once a bin was filled, it was covered over with a generous layer of bagasse and then allowed to compost *undisturbed* for a full nine months or more of "retention time." Undisturbed means absolutely no turning, digging, prodding, agitating, or messing with the compost. The compost is contained, enclosed in bagasse, then covered with bagasse and left alone. This technique eliminates the open compost piles and windrows so commonly used in compost communities worldwide. It also eliminates the need to turn the piles, as there are no exposed surfaces to cool down prematurely, to emanate odors, attract flies, and breed maggots. This "contained static pile" technique also eliminates the tremendous amount of work needed to "turn" hundreds of tons of composting material, by hand. It also eliminates the emission of gases, spores, fumes, odors, and other discharges that result from disturbing active compost piles. The technique also makes the practice of composting on a large scale simple enough to be within the reach of people of meager means.

The temperatures of the piles were monitored and recorded at various locations in the pile. Data showed that the average temperatures remained above 131°F (55°C) for three months or more, significantly above the USEPA-required three-day time period, or the DINEPA-required 50°C for one week (Direction Nationale de l'Eau Potable et de l'Assoinissement in Haiti). The Santo Village temperature results are typical of compost toilet operations as documented by the author, and others, at other Haitian sites and elsewhere in the world.

Prior to adding any organic material to a new compost bin, some dirt in the bottom of the bin is scraped up and piled along the sidewalls of the bin, inside the bin. This gives the bin a shallow bowl-shaped configuration that prevents excess liquid, should there be any, from seeping out the side edges. Instead, excess liquid will remain confined underneath the compost pile. Most of the liquid is absorbed by the compost. The notion that a compost pile is simply a pile of excrement draining into the ground is a misconception often held by people unfamiliar with composting.

Then, a "biological sponge" is constructed in the bottom of the bin. At Santo, a thick layer of bagasse, approximately 1/2 meter deep, was used, although the material could be weeds, leaves, straw, grasses, hay, or other available organic material. This absorbs initial liquids and provides a clean organic layer for the initial toilet material to bed into.

The compost workers were equipped with long rubber gloves, high rubber boots, and coveralls. The collected toilet containers were all dumped into a depression, initially in the biological sponge, but eventually in the compost itself, then covered with clean bagasse or with the existing compost, the final covering being clean bagasse. The bagasse covering prevents odors from escaping, keeps flies from becoming attracted to the compost, and acts as protection from heavy rains and from drying sun.

In effect, the organic material becomes center-loaded into the bins — first a depression is dug, then a layer of toilet material (itself about two-thirds bagasse), then a layer of clean bagasse. The next time toilet receptacles are emptied, another depression is dug, then more toilet material, food scraps, and animal mortalities, if available, more bagasse, and so on, until the bin is heaped full and covered with clean bagasse. The contents will shrink to about half the original volume, or less, after the compost has fully aged, which is why the bins should be filled to maximum capacity.

When each dumping occurs, the bagasse cover is raked to the sides and the fresh toilet material added into the depression. This causes the creation of a bagasse envelope around the compost and prevents toilet material from falling through the openings in the pallet side walls. It also insulates the sides of the compost and eliminates the exposed surfaces that are characteristic of windrow composting. Since there are no exposed surfaces, there is no need to turn the compost piles, as is necessary when open piles, or windrows, are created. High temperatures can be confirmed along the edges of the compost using a compost thermometer.

Windrows and open, uncovered piles have a high surface area to volume ratio. The center of the pile gets hot, but the outside surface does not. For the entire mass of an open pile to be subjected to the internal, biological heat, it must be completely stirred up, several times. The outer surfaces must be turned into the center, a process that is very labor intensive. Open compost piles with large, exposed surfaces also tend to be odorous and covered in flies that are laying eggs and breeding maggots.

After emptying, the toilet receptacles are given an initial rinse with water by the compost management team. This water can be dumped from receptacle to receptacle as they are being scrubbed with a long-handled brush. The initial rinse water, or "black water," is dumped into the active compost bin (the same bin the toilet receptacles are being emptied into). The receptacles are then rinsed again with water. This water is also dumped into the active compost bin. Then the receptacles are sprayed inside and out with clean water, a soap solution, or a disinfectant solution such as bleach, possibly using a hand-pumped sprayer. The clean or soapy water can also be dumped into the compost bin. Water with disinfectant, such as bleach, should be dumped into a soak pit, such as a gravel or stone filled hole. The cleaned receptacles are stored on site until further use. They look and smell clean inside.

The finished compost at Santo Village was independently lab tested at Soil Control Lab in Watsonville, CA, USA. It was considered mature, very stable, safe (tested for fecal coliform, salmonella, and heavy metals), average in nutrients, and a low nitrogen provider with high lime content and an average nutrient release rate, a neutral nitrogen demand, and high ash content. It sprouted healthy test plants. Finished compost can be marketed and sold, bartered, or used by the community for food production.

It is interesting that unsubstantiated fears and concerns caused the compost toilet system to be initially rejected by developers from flush-toilet cultures. This is not uncommon when the composting of human excrement is considered by persons with no experience in the process. Many believe that the process is unsanitary, odorous, and fly-infested and exposes the toilet user to direct contact with human excrement, none of which is true.

The foreign developers had stated that "the social acceptability of this system would probably be quite low as there is repeated contact with the humanure." Yet there was no contact with humanure other than what can be normally expected when anyone uses any toilet. On the other hand, the ELTDT system required the user to periodically remove feces-coated, heavy, fly-infested, 90-liter drums of raw, leaking excrement and dispose of them, without any instruction or guidance for disposal. The undesirability of "having to store humanure for collection" was a nonissue with the compost toilet system. The toilets were emptied and cleaned twice weekly by trained personnel. The dry toilets, on the other hand, were expected to store excrement for up to six months while breeding swarms of flies and emitting intolerable odors that fouled the air of the entire village, even disturbing people in their sleep.

It is unfortunate that gross prejudices and misunderstanding about composting persist in some sanitation and urban planning communities. The ELTDT system, although perhaps suitable for Namibia, collected "waste" for disposal, cost half a million dollars, and failed miserably. On the other hand, when properly managed and carried through to fruition, compost toilets can produce real monetary profit from the sale and use of the finished compost and thereby can create a sanitation system that pays for itself. It can also create local business opportunities such as compost toilet construction, bin construction, gardening, horticulture, landscaping, compost sales, and farming. This is a case study that not only demonstrates how a village-wide compost sanitation system can be organized and implemented, but it also demonstrates how poor planning, prejudice, ignorance, and irrational fear about composting can allow a winning system to be rejected for a losing one.

Data from Santo Village

(Thanks to Alisa Keesey, Patricia Arquette, and GiveLove.org)

Participants

- Households: 250–280 (averaging 4–5 people or 1,000–1,400 persons)
- Average number of households participating per day: 170–200

Compost Bins

- Recycled shipping pallets on edge, fastened together
- Bins needed for 270 households: 32
- Area needed to build bins with 1.5-meter walk between them: 290 m^2
- Dimension of compost bins: 2.4m × 3.2m × 1.2m
- Internal volume of compost bins: 8 cubic meters
- Capacity of compost bins when full: 10 cubic meters (piled high)
- Compost in bin at maturation (after shrinkage): 4–5 cubic meters

Quantity of Organic Material Collected

- Approximately four hundred 5-gallon toilet receptacles are collected each week (averaged over the life of the project), each approximately 2/3 full = 1,333.33 gallons toilet material collected/week = 5.047 cubic meters toilet material collected/week per 250 households.
- 20.19 cubic meters of material collected/month for 250 households.
- Collection is done twice per week.

Cover Material

- 2 cubic yards are needed for 1 cubic yard of toilet material collected.
- The free bagasse is sourced from 2 km away at a sugar factory.
- 13.46 cubic meters of toilet cover material are needed/month for 250 household toilets.
- Assuming 250 households are using the toilets, then each household is using .054 cubic meter/month of toilet cover material ≈ 14.26 gallons/month, rounded to 15 gallons per household per month for cover material utilized at the home (2 qts/day, or about 2 liters/day).
- Assuming 13.46 cubic meters are needed per month for the household toilets, or 161.52 cubic meters per year, but we double that quantity to make sure we have enough for the bins (biological sponge, top cover, etc.), then we need 323 cubic meters per year for 250 households, or approximately 1.3 cubic meters of cover material per household/year.

Compost Management

- One team leader and five compost workers (four are women and two are men) work four hours each per week total (three persons at each compost site), or twenty-four "man hours" per week total, to compost and clean approximately 300–400 toilet receptacles per week, total, at two sites. It takes less than a minute to clean each 20-liter toilet receptacle.
- Composting is done every Tuesday and Friday, at both sites. Households

drop off full toilet receptacles and retrieve clean ones and bagasse as needed.

- The 5-gallon (20-liter) receptacles are about 75 percent full when delivered.
- Composting is done between 5 and 6 am because of the heat.
- The first rinsing of receptacles is deposited into a compost bin.
- Final rinse is with mild bleach solution (to disinfect the outside of toilet receptacles and handles) and is dumped into soak pit
- Toilet receptacles are dried in the sun, then set underneath a tarp.
- Total training time for compost workers was about eight weeks, with a pilot group of twenty-four houses. After the community accepted the system by voting, the project was scaled up from twenty-four homes to three hundred homes over the course of one week. In a survey, 80 percent of households said they knew this system protected the groundwater as well as their drinking water, which they collected on site.

Compost Harvest

- Compost harvested/bin: 1,350 lbs. (612 kg), or 2.5 cubic meters
- Bins harvested per year: 25 (33,750 lbs. or 15,308 kg)
- Compost harvested per year: 62.5 cubic meters or (1,125) 30 lb. bags
- Tons of compost harvested per year: approx. 17
- Metric tonnes of compost harvested per year: 15.42

Compost toilets at the Villa Japon school in Nicaragua replace a pit latrine that was dangerously close to polluting the school water well. The toilets are constructed and maintained by a women's cooperative, which also manages the compost bins and conducts the community education programs when compost toilets are introduced.

Chapter Fifteen

NICARAGUA: SCHOOL AND VILLAGE

GiveLove.org Project: Case Study Review

In the Tipitapa area of Nicaragua, Central America, north of the capital city Managua, a women's cooperative called "Sweet Progress" developed a compost toilet enterprise that provided compost toilets, compost bins, cover material, and compost training to the local community, including the local Villa Japon school, a school of three hundred children. The school's pit latrine "smelled very bad," and the children would rather defecate outside than go inside the stinky, fly-filled room. The manufacturing of the compost toilets and the village-wide expansion of the compost system created numerous jobs for the community. The cooperative also raises honeybees and collects honey, hence their name.

The people understood that the pit latrines were causing groundwater pollution and needed to be eliminated. But their income averaged only about \$2 US per day, so if they didn't use a pit latrine, they simply resorted

Continued on page 219

The new Villa Japon school compost toilet building replaced the blue pit latrine building in the background, which was sitting dangerously close to the school water well.

GiveLove
Sweet
Progress

Photos these two pages contributed by Sweet Progress.

The Villa Japon school in Nicaragua accommodated three hundred children. What you are looking at here is nineteen months of the school's toilet material collected in these compost bins. Otherwise, it would have gone into a hole in the ground dangerously close to the school's water well. The ladies who managed the system laid thorny branches on top of the bin to ward off horses that wanted to eat the grassy cover material. The children first learned how to use the compost toilets; then they took their knowledge home to their parents, who subsequently converted from pit latrines and open defecation to compost toilets, which greatly improved their quality of life.

to open defecation. Although most homes had water piped to an outdoor faucet, the water was only available half an hour a day and was not available on many days.

Sweet Progress manufactured the compost toilets from local and recycled materials, including repurposed laminated cardboard, costing only about $25 US each (2016), including the compost bins and community training. Wood toilets cost about $50 US each, also including the bins and training. One pit latrine, on the other hand, cost ten times as much as an entire compost toilet system, which is probably why so many people in the community practiced open defecation — it doesn't cost anything. The money for the compost toilets was being donated by GiveLove.org.

The primary cover material used in the toilets was rice hulls, or husks, from a local rice factory, free for the hauling. The husks were stored outside and exposed to the rain to allow for hydration and increased biological activity. Rice husks tend to be difficult to break down in compost, and although they will work as cover material, they can reduce the temperatures in the compost piles unless additional green materials or food scraps are included in the mix.

This 70-year-old lady built her own compost bin in Nicaragua.

Compost toilets require little space and can be located anywhere that is convenient for the user. They should be kept in a private, comfortable, secure space where a person can go about his or her business in a relaxed and happy manner. A private, odor-free, pleasant indoor toilet should be available to every human being. Compost toilets provide a simple, low-cost, easily managed option where water toilets are impossible, or where compost is preferred to sewage. Unlike pit latrines, compost toilets are safe for children, who can use them unattended without the fear of a deadly fall into a pit of sewage teeming with maggots.

Photo opposite page contributed by Sweet Progress.

Sugarcane bagasse, which is used in many tropical areas, unlike rice hulls, contains sugar residues which help feed bacteria. Bagasse may have an *inorganic* silica content of approximately 9.78 percent while rice husks may contain up to 20 percent by weight of silica. The higher inorganic component of rice husks may be one reason rice husks seem slower to compost than bagasse. The compost piles at the Villa Japon school were only reaching temperatures of about 118°F (48°C) when monitored by the author, whereas bagasse/toilet material compost in other countries typically exceeded 131°F (55°C) and was often much higher. Lower internal compost bin temperatures require longer retention times for adequate pathogen reduction and agricultural conversion before the compost can be used.

The compost bins were primarily made from wood pallets, a by-product of the shipping industry that were readily available at a low cost. They can easily be assembled into a quick compost bin by setting the pallets on edge and fastening them together, by tying, nailing, or screwing. Once the pallets are available on-site, a family-size bin can be constructed in minutes. At the Villa Japon school, the compost managers simply laid thorny branches on top of the compost piles to prevent horses from eating the grassy layer covering the compost.

In Tipitapa, the compost toilets were equipped with 20-liter plastic receptacles which lifted out of a hinged toilet cabinet. The toilets were provided to twenty-five households during the first phase of the project. An eighty-year-old man living in a small, crude dwelling put his compost toilet immediately next to his bed for convenience. It was the first time in his eight decades of life that he had had an indoor toilet.

Before the families had these toilets, they had either engaged in open defecation or used pit latrines. The pit latrine holes were a danger to small children, who can fall in. Bad odors, now nonexistent, had been abundant. One seventy-year-old lady who had converted from open defecation to a compost toilet was asked if the compost toilet system was hard to take care of. She had built her own compost bin, and she immediately replied, "No." She remarked that her compost toilet, located under a roof attached to her dwelling, was "like an expensive flush toilet in a big city." She and her daughter both agreed that the toilet had improved their lives.

Chapter Sixteen

MOZAMBIQUE
COMPOSTING BLACK WATER
Case Study Review

Yes, you can compost liquids. People who say that you shouldn't include urine in compost toilets simply don't know what they're talking about. You can collect nothing but urine in a "compost urinal" and compost it later in a compost bin by filling the urinal with a carbon-based cover material such as sawdust or bagasse, starting with a few inches in the bottom, then sprinkling in a little more after each urination. Once filled, add it all to a compost bin. The author has done so for decades. So when asked to show people how they could compost black water from school toilets in eastern Africa in 2013, he agreed.

In this case, public schools in Mozambique, Africa, had been equipped with concrete toilet buildings. These collected toilet material under the floor in a vault similar to a shallow swimming pool. There was no cover material, and nothing was added to the toilet contents. Whoever came up with this design probably was not thinking about what to do when the system filled up. Not to mention the odors, flies, and maggots.

It turned out that the toilet material was almost all liquid. It was surmised that most of the children defecated at home either before or after school and used the school latrine primarily for urination. Equally possible is that the solids simply dissolved into the liquids, yielding a "black water" that could be pumped out by a pump truck. What to do with the black water? That was the question.

The location was Xai-Xai, Mozambique (pronounced *Shy Shy*), the economic capital of the province of Gaza, located in the southern end of the country about 200 kilometers north of the capital city, Maputo. According to the 2007 census, Xai-Xai's population was 104,689, consisting of about 23,000 households.

The public schools' toilet liquids were going to be pumped out of the toilet buildings and then discharged into a "drying bed" located outside

School toilet buildings had a concrete vault underneath the floor that could be pumped out via a rear hatch, shown below. The vault filled with black water, which needed to be ecologically processed. A good source of sugarcane bagasse provided a solution: soak the bagasse with the black water and cocompost it with market scraps.

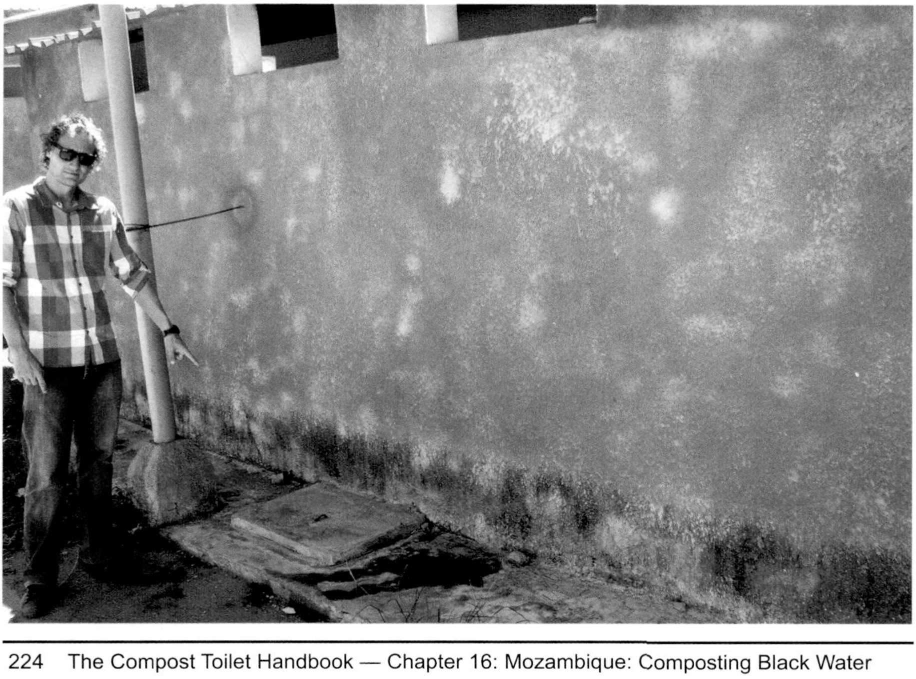

of fluid can be estimated based on the absorption rate experiment in the barrel. Then the mixture would be left to sit for a period of time. At this point, the valve is closed so no liquid can drain out of the absorption bed. After waiting, the valve is opened, and any excess liquid is drained out into the leachate tank. At this point a compost thermometer could be inserted into the bed to see if the contents start to heat up inside the absorption bed. During this time the valve remains open so the bagasse can continue to drain. Excess fluid can be removed from the leachate tank into drums to prevent backflow into the bed, if necessary. The purpose of this experiment would be to see if thermophilic conditions will begin inside the absorption bed. If so, then the material should be left inside the bed for a couple of weeks to reduce or eliminate the pathogen load. Then the organic material can be removed from the bed and placed into compost bins either alone, or with green market material.

Wood compost bins were constructed for composting our black water/bagasse combination. Food scraps from the town market were also collected and added to the mix in the bins, allowing for a more robust compost. The yellow plastic was being used to line toilet vaults. Here it lines a compost bin on an experimental basis.

The toilet liquid could contain intestinal parasites, or eggs, and even bacteria such as the cholera pathogen. Therefore, it should be subjected to a composting process to eliminate the pathogens and return the organic material to the soil in a hygienically safe manner.

Handling toilet material is unsanitary and exposes workers to health hazards. If black water can simply be pumped into an absorption bed, soaked into bagasse, drained, and then achieve thermophilic conditions inside the bed, the process will be much more sanitary. Otherwise, workers will need to remove the soaked bagasse from the bed and pile it into com-

Workers experiment with the absorption rates of bagasse and toilet water. Note the concrete leachate tank, which collects liquids draining from the drying bed.

post bins. If the mixing tank process works, it could have a wide application anywhere in the world where toilet fluid is a problem and where carbon material is available.

A well-established alternative would be to collect toilet material in containers under the toilet seats, cover the toilet contents with bagasse, sawdust, rice hulls, or ground plant material after each use, then compost it in separate bins. This is the process we have seen throughout this book. Not only does this prevent odors and flies at the toilet, but it mixes the toilet contents with the carbon material at the point of collection. The toilet material would be composted near the toilets. This would eliminate the pumping, transport, and handling of unsanitary material. Once the compost has fully matured, it becomes greatly reduced in mass and is biologically stable, odorless, not attractive to vermin, and indefinitely storable. Then it could be either hauled off the site or left for use on the school grounds, as demonstrated at the Amurt school (and others) in Haiti. An advantage to this approach is that it could provide a learning experience for the school children. They would have a unique opportunity to learn about composting, recycling, health, hygiene, gardening, sustainability, and agriculture all rolled into one living laboratory on the school grounds.

Bagasse or sawdust would have to be delivered to the school grounds and a stockpile always maintained there for such a toilet system to work. If there is a market for the compost, a private enterprise could provide the cover material, then manage and collect the finished compost for its own processing, sale, and profit.

The sugar factory had thousands of workers in villages with no sanitation, an unlimited supply of bagasse, money to finance a compost sanitation system, and the desire to provide some form of sustainable sanitation for its employees. If they set up compost sanitation systems at these villages, a lot of good data could be collected that may make it easier to establish large-scale compost sanitation in schools, villages, refugee camps, and other such places elsewhere.

The composting of toilet liquid and market scraps in Xai-Xai appeared to be feasible if a carbon-based material is available in large enough quantities to combine with the nitrogenous, wet input from the toilets. This

Reed walls surround a local toilet that has a sunken vault, shown below. The vault is lined with a yellow plastic material, then pumped out when full. The material can be composted, although it was being discharged at the municipal dump instead.

project provided an excellent and unique opportunity, a living laboratory, where liquid sewage can be recycled via composting. A parallel toilet system where the toilet contents are collected under the toilets, with bagasse, and composted in bins near the toilet stalls, could be established at schools, at homes, or at the sugarcane community, and the two systems (pump truck and direct compost) compared side by side for an even more enriching research project.

> From the author's Field Notes: *"When riding up to Xai-Xai from Maputo on the first day, the project boss rode with me. I told him that I like to teach local people and that the worst ones to teach are the European and especially American whites, particularly those with advanced university degrees. They have to "unlearn" before they can learn new things. You can't pour into a glass when it's already full, I told him. The whites believe that defecating into pots of drinkable water, then flushing it away, is the highest form of sanitation, and that anything else is beneath that. Whereas the locals are used to defecating in holes in the ground. When they see a scientifically proven alternative that eliminates flies and odors and is ecologically sound, even helping their gardens, they show an interest. They follow instructions and do what I tell them, and it works."*

It has always been a pleasure working with the local people. They are receptive, interested, and intelligent, and they learn quickly and follow directions well. The most successful compost sanitation projects are those managed by the toilet users, the same ones who make and use the compost. They are greatly improving their sanitation situation, providing comfortable toilets for themselves or their communities, and capturing soil fertility that had always been otherwise lost. They take pride in their work and are highly respected for the work they do, and rightly so.

It may be difficult to imagine that many people can congregate for long periods of time, months, or years, even generations, and produce gardens instead of sewage. The Kailash Ecovillage provides a living example where such a phenomenon is not only possible but demonstrated daily. When the electricity goes out, the ecovillage will still have functioning toilets, unlike their neighbors who are connected to a central sewage system relying on electricity, pipes, pumps, drains, and running water.

Photos for Chapter 17 provided by Ole Ersson.

Chapter Seventeen

USA ECOVILLAGE

All the case reviews in this book so far have been in developing countries where water toilets are rare, if not impossible. Yet there are many compost sanitation examples in North America, Australia, Europe, and elsewhere in the developed world where flush toilets are the norm.

Developed, highly populated urban areas may impose restrictions that are not required in underdeveloped rural societies, such as a waterproof barrier between the compost and the soil and other infrastructure regulations that can greatly add to the expense of a compost system. However, where more money is available, a more elaborate system can be created. One such example is the Kailash Ecovillage in Portland, Oregon, USA.

Many centralized sewer systems in the developed world are extremely expensive and wasteful while inadequately eliminating pollutants and pathogens before discharging treated wastewater back into the natural environment. They are also subject to catastrophic failures during natural disasters such as earthquakes, hurricanes, wildfires, or floods.

On the other hand, decentralized compost sanitation systems don't waste energy, don't pollute potable water, and don't squander soil fertility.

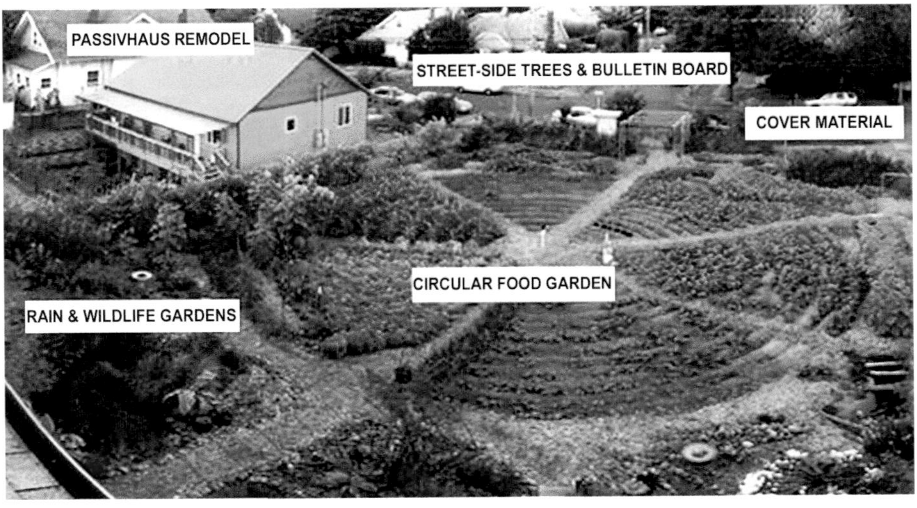

Kailash Ecovillage

Instead, they create agricultural nutrients for growing food. Compost toilets will still work when the water stops running, the electricity stops flowing, and your money runs out, so long as there is a supply of cover material and a compost bin for processing the organic material. After the tornado, the earthquake, the hurricane, the flood, or the wildfire, you can have your compost toilet up and running again in no time.

Ecological sanitation, or *ecosan* for short, continues to be demonstrated around the world. The Kailash Ecovillage provides a prototype for an

Compost Toilets at Kailash Ecovillage

urban system that has been successful for years. The ecovillage compost system requires durable, corrosion-resistant materials while allowing no discharge of unprocessed material, such as compost leachate, into the environment. It utilizes enclosed, ventilated, vermin-proof compost bins. The finished compost is tested to insure pathogen elimination. Collected urine is sequestered for six months to ensure pathogen elimination before use as a fertilizer. For the benefit of the users, an operation and maintenance manual is provided that includes clear instructions.

The multichambered compost bin is constructed with a waterproof concrete pad that has a 4-inch-high lip around its edges, channeling any leachate into a central drain.

The ecovillage collected four years of data from two passes through all of the compost bins. Pathogen test results exceeded the USEPA (United States Environmental Protection Agency) "Class A" compost standards. To quote the EPA in relation to sewage sludge (a by-product of wastewater treatment plants), *"Composting is the biological degradation of organic materials in sewage sludge under controlled aerobic conditions. The process is used to stabilize wastewater solids to create a marketable end-product that is easy to handle, store, and use as a soil amendment. The end product is often a Class A, humus-like material without detectable levels of pathogens, that can be applied as a soil conditioner and fertilizer to gardens, food and feed crops, and rangelands."*

Kailash Ecovillage's urban compost sanitation design offers one of the simplest, most robust, and workable demonstration systems available.

For more information:

https://iwaponline.com/bgs/article/1/1/33/69034/The-Kailash-Ecovillage-project-converting-human

The compost bins have a concrete floor with a built-in leachate collection system.

CAN'T FLUSH?

It's just another day in the life of a turd.

Chapter Eighteen

EMERGENCY PREPAREDNESS

Compost toilets operate without electricity or running water and therefore are reliable when other toilet systems are not. As the twenty-first century grinds on, we seem to be finding ourselves confronted with more and more natural disasters: hurricanes, earthquakes, tornadoes, floods, wildfires, ice storms, windstorms, and so on. Add homelessness, temporary encampments, refugee camps, temporary military outposts, "it" hitting the fan, and the like, and it should be obvious that we ought to be taking a look at sanitation systems that are reliable without needing electricity or running water; systems that are better than the "collect and dump" portable chemical toilets currently in widespread use in many parts of the world.

Compost toilet systems can be rapidly scaled up or down, depending on the need, even during disasters and even in large populations. It seems wise to incorporate such systems into disaster preparedness and response plans. It would help if regulators supported the permitting of compost sanitation systems and if they developed regulations, policies, and guidelines that encouraged compost-based sanitation.

The United States has over seventeen thousand wastewater treatment plants serving three-quarters of the population. These plants don't work without electricity and running water. What happens when *your* electricity suddenly shuts off and you have no idea when, or if, it will come back on? Well, what happens is that you begin to have some idea of what it's like for the billions of humans who live without electricity or running water every day. Suddenly, you don't have a functioning water toilet. What do you do?

The American Society of Civil Engineers rates American wastewater infrastructure as a D+, slightly above a failing grade. They estimate that fifty-six million new users will be connected to American wastewater facilities by 2032. That's a lot of people with flush toilets in the US alone, and a lot of toilet material that will continue to be produced by their bodies whether they can flush their toilets or not. It is estimated that the US will need $271 billion to meet current and future wastewater treatment needs.

Although this sounds like a huge amount of money, it is equivalent to the amount US taxpayers spend on military expenses *every 14 weeks*. It seems ironic, and unfortunate, that spending money to protect the public from large-scale sewage pollution in the event of a disaster is not considered "defense" spending, and therefore "defense" money can't be allocated for these purposes.

After a hurricane or other extreme wind or weather event, trees are downed, possibly by the thousands. In Louisiana, USA, after Hurricane Katrina, trees were laid flat over a 50-mile (80 km) expanse. Many trees were blocking roadways, breaking electric lines, damaging houses, and so on. These trees needed to be removed and cleaned up. The branches could have provided the potential for many tons of fresh cover material for emergency compost toilets if ground up fine enough to use for this purpose. In a disaster scenario, we can clean up the trees and debris and prevent sewage at the same time by using composting as a sanitation option. Put in perspective, a US military air-to-air missile costs upwards of a half million US dollars. How many chipper/shredders or compost grinders would that buy?

A group of three hundred people gathered in Kentucky, USA, for a music festival in a remote wooded area where toilets had to be provided. Portable chemical toilets and pump trucks were not practical due to the rugged terrain and lack of access. Instead, heavy-duty plastic, wheeled "garbage cans," also known as "wheelie bins" were used to collect the toilet material. The approximately 50-gallon (190 liter) bins were positioned directly underneath the toilets. A cover material was used after each deposit. When about 75 percent full, the bins were wheeled out from underneath the toilet building and replaced with empty bins. The "full" bins were set aside with lids.

About three hundred people filled six bins over the duration of the festival. What is interesting about this scenario is that nothing was done with the bin contents. The full bins just sat there for an entire year until the author came upon them a year later while conducting research. What he found was that the bin contents had shrunk by at least half. There was almost no odor inside the bins. He dumped the bin contents into a pallet bin set up for making compost. New organic materials, including toilet material and food scraps, were then added to the pallet bin as needed.

What does this have to do with disaster preparedness? Clearly, if a family in an urban situation had a compost toilet available to them and had a bale of compressed peat moss, or a bag of rice husks, sawdust, or fine wood shavings, and a couple of wheelie bins, they could have a quick, functional sanitation system that could be used for months. The toilet contents could be deposited into the wheelie bins, as could food scraps. Once the disaster scenario had passed, the wheelie bins could be collected by a municipal authority and composted off-site later. It won't hurt the organic material in the bins if many months pass before the bins are emptied.

The top photo shows six wheelie bins that collected toilet material from three hundred people over several days. The bins still had the original material inside a year later. The contents shrank to less than half its original volume. The lower photo shows bins in position underneath the toilet building, collecting toilet material directly deposited by users.

An emergency toilet need be nothing more than a toilet receptacle, snap-on toilet seat, cover material, bulk storage bin, and some pertinent reading material, of course.

Alternatively, centralized community compost bins could be constructed, and the wheelie bin contents collected there, managed by trained personnel. Finally, individual compost bins quickly constructed from pallets, for example, could be used for collecting the toilet material in a suburban backyard, provided an adequate supply of suitable cover material is available, and the users had some composting knowledge or training.

This would allow continuous use of comfortable, odorless, secure indoor toilets without the need for water or electricity. A small amount of water would be needed to wash out the toilet receptacles, or else compostable plastic bag liners could be used in the receptacles, thereby eliminating the need for water altogether (as was done at Standing Rock). Approximately one plastic bag liner per person per week would be needed.

This emergency sanitation system can be scaled up to include large, waterproof "dumpsters" where toilet material could be collected by emergency sanitation workers. The contents of the dumpsters could be composted once circumstances allow. The object is to keep all toilet material from entering the environment. If you can't flush it, that doesn't mean you're justified in dumping it outside or into a hole in the ground. Use a carbon-based organic cover material and collect it all in a watertight container so that it can be composted at a later date.

Emergency preparedness supplies would include a compost toilet with enough receptacles (with lids) to comfortably service the number of people using it. The toilet can be as simple as a 5-gallon (20-liter) plastic receptacle with a snap-on toilet seat. These tend to be unstable on smooth surfaces where they can tip over when you're sitting on them and leaning sideways (to wipe), so it's better if the toilet is enclosed in a stable cabinet of some sort. One adult is going to fill one receptacle per week, depending on the size of the adult, how much he or she eats, and the quality of the cover material being used. Loose, fluffy cover material such as wood shavings will fill a receptacle much faster than finer material such as sawdust, especially if the sawdust has some moisture content, which greatly improves the "biofilter" qualities, or odor-blocking characteristics of the sawdust.

Cover materials can be bought in loose bags or in compressed form at feedstores or garden supply outlets. These can include peat moss, fine wood

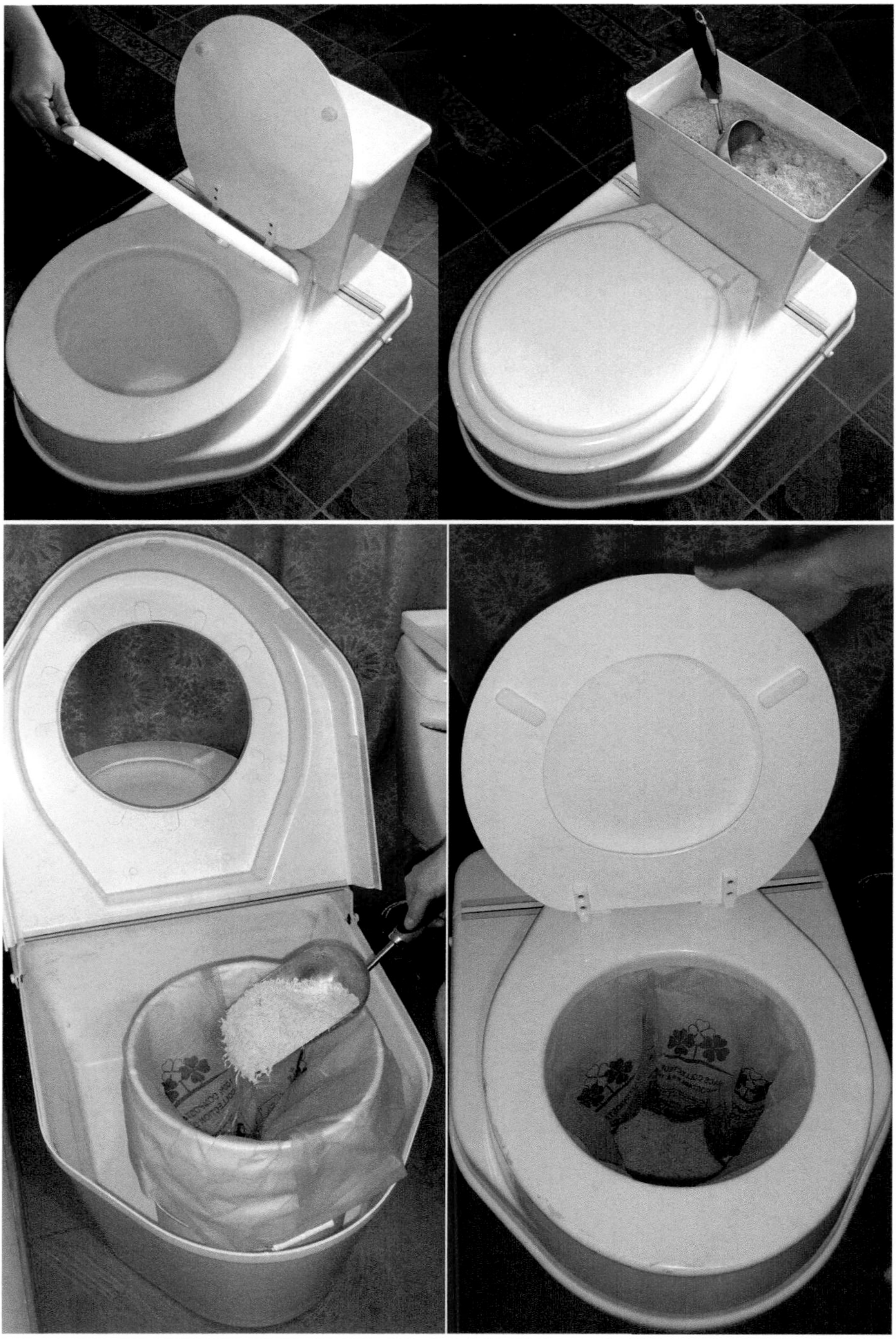

Compostable plastic bag toilet receptacle liners can reduce or eliminate the need for cleaning water in emergency situations. A 13-gallon bag is used in 5-gallon receptacles.

shavings, and rice hulls. The shavings and husks are typically used for animal bedding. The moss is typically used for soil mixes. These will all be dry and should be rehydrated somehow for optimum performance. They can be left outside in the rain to rehydrate and become biologically active, or they can be misted with water before adding to the toilet. For urinals, just leave the material dry.

Have watertight 50-gallon wheelie bins, with lids, available. Include several boxes of 13-gallon compostable plastic bags as toilet liners, ideally a brand that is suitable for backyard compost (it should say so on the box). These are readily available for purchasing on the internet. First, put about 12 inches of dry cover material in the bottom of the wheelie bins, then dump the toilet contents, bags and all, into the bins. There is no need to puncture the bags beforehand. When the bin is full, cover the contents thoroughly with clean cover material, close the lid, and deal with it later when circumstances allow.

Finally, keep an emergency supply of toilet paper on hand.

[Incidentally, although the author "invented" the Loveable Loo compost toilet and trademarked the name, he allowed the trademark to lapse in 2020. It is now an open-source concept and not a source of remuneration for the author. Also, *The Humanure Handbook* has been available for free download since the year 2000, including the most recent fourth edition published in 2019. The point being that the mentioning of the Loveable Loo and *The Humanure Handbook* in this book is not meant to be misconstrued as a commercial gesture. The meager income generated by the author from his work in compost sanitation is used to help finance his travel, research, and publishing. All financial support is appreciated (such as paying for books) but is not a primary concern of the author.]

Special thanks to Ryan A. Smith for his September 2020 postgraduate thesis entitled *Disasters, Finances, Nutrients, and Climate Change: A Case for Waterless Sanitation Systems* (Naval Postgraduate School, Monterey, CA), which provided the inspiration for this chapter.

COMPOST TOILET REVIEW

DO — Collect urine, feces, and toilet paper in the same toilet receptacle. Urine provides essential moisture and nitrogen.

DO — Keep a supply of clean, organic cover material handy to the toilet at all times. Rotting sawdust, peat moss, leaf mould, and other such cover materials prevent odor, absorb excess moisture, and balance the C/N ratio.

DON'T — Worry about your compost. If it does not heat to your satisfaction, let it age for a prolonged period, then use it for horticultural purposes.

DON'T — Segregate urine or toilet paper from feces.

DON'T — Turn the compost pile. Sit back and relax and let the microbes do the work for you. Once your pile is built, let it age in peace, undisturbed, for approximately a year.

DON'T — Use lime or wood ashes on the compost pile. Put these things directly on the soil.

DO — Keep another supply of cover material handy to the compost bins for covering the compost pile itself. Coarser materials, such as hay, straw, weeds, leaves, and grass clippings, prevent odor, trap air in the pile, and balance the C/N ratio.

DO — Deposit humanure into a depression in the top center of the compost pile, not around edges.

DO — Add a mix of organic materials to the humanure compost pile, including food scraps.

DO — Keep the top of the compost pile somewhat flat. This allows the compost to absorb rainwater and makes it easy to cover fresh material added to the pile.

DO — Use a compost thermometer to check for thermophilic activity.

DON'T — Expect thermophilic activity until a sufficient mass has accumulated.

DON'T — Deposit anything smelly into a toilet or onto a compost pile without covering it with a clean cover material.

DON'T—Allow dogs or other animals to disturb your compost pile. If you have problems with animals, install wire mesh or other suitable barriers around your compost, and underneath, if necessary.

DON'T — Segregate food items from your humanure compost pile. Add all organic materials to the same compost bin.

DON'T — Use the compost before it has fully aged. This means one year after the pile has been constructed, or two years if the humanure originated from a diseased population.

Girl photos by Samuel Autran Duorada at the Kwale's girl's high school, Kwale, Kenya.
(an Aqua-Aero WaterSystems project)

INDEX

Bold indicates a primary reference.

D

E

F

G

H

I

J

K

L

M

S

T

U

V

W

X

Y

Z

COMPOST TOILET SITES IN THIS BOOK

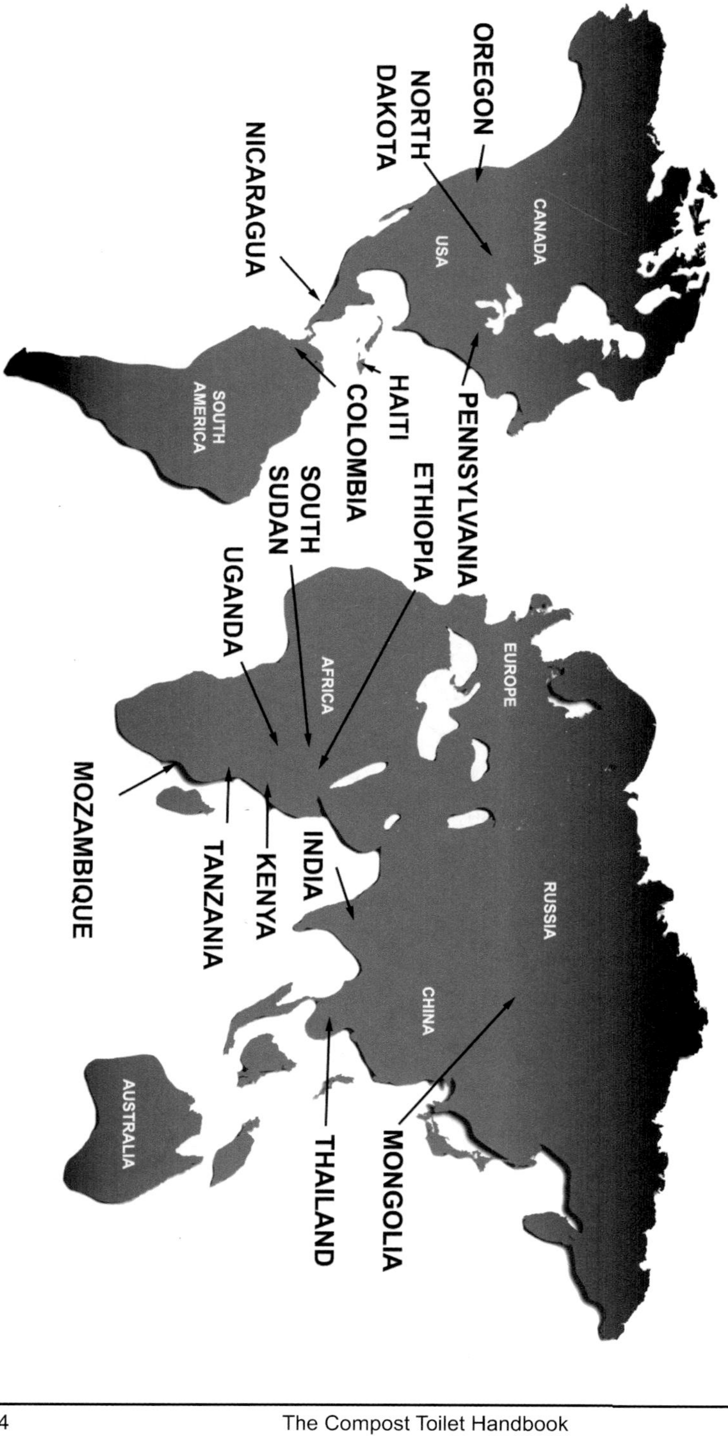